TOWARDS SUSTAINABLE ARTIFICIAL INTELLIGENCE

A FRAMEWORK TO CREATE VALUE AND UNDERSTAND RISK

Ghislain Landry Tsafack Chetsa

Apress®

Towards Sustainable Artificial Intelligence: A Framework to Create Value and Understand Risk

Ghislain Landry Tsafack Chetsa
London, UK

ISBN-13 (pbk): 978-1-4842-7213-8   ISBN-13 (electronic): 978-1-4842-7214-5
https://doi.org/10.1007/978-1-4842-7214-5

Managing Director, Apress Media LLC: Welmoed Spahr
Acquisitions Editor: Shiva Ramachandran
Development Editor: Matthew Moodie
Coordinating Editor: Nancy Chen, Rita Fernando

Cover designed by eStudioCalamar

Cover image designed by Freepik (www.freepik.com)

Distributed to the book trade worldwide by Springer Science+Business Media New York, 1 New York Plaza, New York, NY 100043. Phone 1-800-SPRINGER, fax (201) 348-4505, e-mail orders-ny@springer-sbm.com, or visit www.springeronline.com. Apress Media, LLC is a California LLC and the sole member (owner) is Springer Science + Business Media Finance Inc (SSBM Finance Inc). SSBM Finance Inc is a **Delaware** corporation.

For information on translations, please e-mail booktranslations@springernature.com; for reprint, paperback, or audio rights, please e-mail bookpermissions@springernature.com.

Apress titles may be purchased in bulk for academic, corporate, or promotional use. eBook versions and licenses are also available for most titles. For more information, reference our Print and eBook Bulk Sales web page at http://www.apress.com/bulk-sales.

Any source code or other supplementary material referenced by the author in this book is available to readers on GitHub via the book's product page, located at www.apress.com/978-1-4842-7213-8. For more detailed information, please visit http://www.apress.com/source-code.

Printed on acid-free paper

To my family Valerie and Adlynne, and my sister Bertine.

Contents

About the Author

Ghislain Landry Tsafack Chetsa is Head of Data Science at Elemental Concept 2016 Ltd (EC), where he steers the organization's AI strategy. As part of this, he leads the company's work in leveraging the latest advances in AI to help clients create value from their data and auditing AI systems developed by third parties on behalf of potential investors.

As part of Ghislain's work in the healthcare industry at EC, he is supporting the development of data-related healthcare products for his clients. This made him appreciate the challenges involved in and the complexity of developing AI systems that people trust to make the right decision for them. It further motivated him to write this book.

Before joining EC, Ghislain held positions as performance engineer and data scientist in the telecommunications and energy sectors. Prior to this, Ghislain worked as an academic at the French National Institute for Research and Automation (INRIA) and the University of Lyon I. His work primarily focused on analyzing the behaviors of high-performance systems to improve their energy efficiency, and this gave him the opportunity to coauthor several scientific publications presenting methodologies for improving the energy efficiency of large-scale computing infrastructures. He holds a PhD in computer science from École Normale Supérieure of Lyon, France.

Acknowledgments

I am immensely grateful to my partner Valerie for her unconditional love, companionship, and endless support, for always being there for me, the edits on this book, and the many valuable discussions. Thank you, Valerie, for being such a great mum to our baby girl.

I am grateful to Glen Moutrie, and Sharva Kant, who accepted to review this book, for their valuable comments on a preliminary version of this book.

I would also like to thank all my colleagues and friends at Elemental Concept for their support and encouragement.

I am grateful for the support and guidance of Shivangi Ramachandran, senior editor for Business and Business Tech, who believed in me and this project. This book would not have been possible without her.

To my coordinating editor, Rita Fernando, I owe many thanks for her hours of work and guidance about many aspects of this book. A special thank you goes to Nancy Chen, whom I started this project with.

And to Matthew Moodie, my development editor, for all the guidance and the many hours he spent making sure that this book is what it is today.

Finally, I also express my gratitude to Elemental Concept 2016 Limited which gave me the opportunity to work on projects that motivated me to write this book.

Introduction

Artificial intelligence (AI) by now infiltrates various aspects of daily life. While some are skeptical of its potential, others believe AI can provide fast and effective approaches for addressing a wide range of problems: from policing and crime prevention to personalized healthcare or streamlining the judiciary system. This is a belief which is likely motivated by the successful application of AI in industries such as the retail and the manufacturing industries. It is, therefore, not surprising that many organizations, including public institutions, are trying to adopt AI and related technologies in some shape or form.

It should, however, not be surprising that adopting a technology that has as much potential as AI generally requires many more inputs than the technology itself. In other words, recruiting data-related professionals or experts and having them develop AI systems for an organization is unlikely to be sufficient to meet all its objectives.

In fact, we observed a similar trend with the advent of the Internet.

With the use of the Internet increasing throughout the 21st century, users have increasingly been subject to abusive practices such as unwanted commercial emails (spam), identity theft, and more recently user tracking. Experts have responded to and mitigated some of these through technical solutions, while regulators and governments have stepped in to prohibit certain practices. In some cases, it is now required that organizations inform their users of their practices in advance. Importantly, we as users expect organizations that we interact with online to mitigate our exposure to abusive practices. This means that from an organization's perspective, simply creating an online presence is probably not the biggest challenge. The real complexity lies in understanding the following questions: What are the greatest potential challenges, and are they addressed adequately? How is the potential return on investment in building an online presence impacted by such challenges? What is the impact of the online presence on my organization's ability to create value in the short, medium, and long term? In essence, organizations need to invest to understand and manage the risk around their online presence.

Investing in AI is not so different: Successfully developing AI systems that meet the constantly changing needs of today's society requires a holistic, methodical, and adaptive approach designed to assess and manage the risks around introducing AI. Such an approach should ensure that AI systems developed by an organization are aligned with its way of conducting business and meet certain social standards. The increasing importance of principles such as fairness, privacy, and social equality in discussions around AI corroborates the importance of this. Failing to meet those standards can result in lost opportunities for an organization. Worse yet, it may even lead to an organization's demise.[1]

Preventing or managing the social, political, and economic challenges that organizations wishing to develop and/or deploy sustainable AI systems may face today is of great importance. We understand sustainable AI systems to be those that manage to incorporate both business and human values and principles in their design. Potential challenges include, for example, an AI system's need to remain consistent with the company's as well as human principles to be truly effective and/or remain flexible to adjust to new regulations imposing alignment with such social standards. While it is often easy to identify a problem once it has happened, making sure that similar problems do not arise in the future is challenging because it requires a methodical approach involving all stakeholders.

This book discusses the concept of sustainability in AI. To help businesses achieve the objective of developing sustainable AI systems, this book introduces the sustainable artificial intelligence framework (SAIF). This is designed to help organizations create value and understand risks when adopting and/or developing AI systems.

Audience

The challenges addressed in this book are of interest not only to mainstream adopters of AI but also to organizations seeking to improve their efficiency through AI.

The primary audience for this book is decision makers in organizations wishing to deploy AI systems. This includes government officials, members of the C-suite or other business managers, data scientists, as well as any technology experts aspiring to a data-related leadership role. While the book refers to the technical aspects of AI in places, sufficient context is provided whenever needed to facilitate understanding. As a result, the reader is not required to have a deep background knowledge of AI to understand the key concepts of the methodology discussed in this book; a basic training in statistical methods should be sufficient. This being said, readers with some knowledge of AI may find the book more accessible.

---

[1]The example of Cambridge Analytica demonstrates this.

What You Will Learn

By the end of the material presented here, the reader will learn

- How to develop and deploy AI systems that meet certain social standards and principles

- How to identify ethical risks in the development of AI systems

- How various aspects of data make it both an asset and a liability, and why today's data may be inadequate for future projects

- How to design and implement an effective AI strategy along with necessary governance, and how to assess the effectiveness of the AI governance strategy

- How to assess the impact of AI on an organization's ability to create value in the short, medium, and long term

- The principles and requirements of effective regulation for AI

Organization

This book is structured as follows:

Chapter 1 provides an overview of the complex relationship between AI and today's society and introduces the four pillars of the sustainable AI framework. Chapter 2 reviews the human factor pillar of AI; it further discusses the relevance of ethics and sources of ethical hazards in AI. The SAIF framework for the development and deployment of AI systems is then introduced in Chapter 3. Chapter 4 highlights the need for an understanding of AI at the organization's level and introduces the concepts of AI governance. Chapter 5 summarizes existing performance metrics used to evaluate AI systems and introduces a new framework to account for the human factor of data science (DS). Chapter 6 presents a practical example of an implementation of the SAIF framework in the form of the design and implementation of an AI-powered credit scoring system. Chapter 7 discusses regulatory mechanisms for AI along with their limitations; it further introduces an effective regulatory arrangement for the development and deployment of AI systems. A discussion around the development and deployment of AI in the medical decision context can be found in Chapter 8. Finally, Chapter 9 presents concluding remarks and a discussion on the need for standards and definitions.

AI in Our Society

Up to the early 2000s, artificial intelligence (AI) was perceived as a utopia outside of the restricted AI research and development community. A reputation that AI owed to its relatively poor performance at the time. In the early 2000s, significant progress had been made in the design and development of microprocessors, leading to computers capable of efficiently executing AI tasks. Additionally, the ubiquity of the Internet had led to data proliferation, characterized by the continuous generation of large volumes of structured and unstructured data at an unprecedented rate. The combination of the increasing computing power and the availability of large data sets stimulated extensive research in the field of AI, which led to successful deployments of the AI technology in various industries. Through such success, AI earned a place in the spotlight as organizations continue to devote significant effort to integrate it as an integral part of their day-to-day operational strategy. However, the complex nature of AI often introduces challenges that organizations must efficiently address to fully realize the potential of the AI technology.

This chapter provides an overview of the increasingly complex relationship between AI and today's society and introduces the foundational concepts of the sustainable AI framework.

© Ghislain Landry Tsafack Chetsa 2021
G. L. Tsafack Chetsa, *Towards Sustainable Artificial Intelligence*,
https://doi.org/10.1007/978-1-4842-7214-5_1

The remainder of the chapter is structured as follows:

- First, we discuss the relevance of AI in today's society.

- We then present various challenges that AI faces today.

- Finally, we introduce the concept of sustainable AI and examine how these challenges can be addressed.

The Need for Artificial Intelligence

Jhon MacCarthy, one of the founding fathers of the artificial intelligence discipline, defined AI in 1955 as

> [...] *the science and engineering of making intelligent machines, especially intelligent computer programs* [...][1]

where

> *intelligence is the computational part of the ability to achieve goals in the world*[1]

MacCarthy further provided an alternative definition or interpretation of AI as

> [...] *making a machine behave in ways that would be called intelligent if a human were so behaving* [...][1]

Over the years, and especially within the business world, the definition or interpretation of the term AI has evolved or has been altered to incorporate development and progress made within the discipline. For example, Accenture in its "Boost Your AIQ" report defines AI as

> [...] *a constellation of technologies that extend human capabilities by sensing, comprehending, acting, and learning — allowing people to do much more*[2]

---

[1]www-formal.stanford.edu/jmc/whatisai/node1.html
[2]www.accenture.com/t20170614T050454_w_/us-en/_acnmedia/Accenture/next-gen-5/event-g20-yea-summit/pdfs/Accenture-Boost-Your-AIQ.pdf

Likewise, PWC's "Bot.Me: A Revolutionary Partnership" Consumer Intelligence Series report defines AI as

> [...] *technologies emerging today that can understand, learn, and then act based on that information*[3]

Other definitions exist: for example, the Oxford dictionary defines AI as

> [...] *the theory and development of computer systems able to perform tasks normally requiring human intelligence such as visual perception, speech recognition, decision making, and translation between languages.* (Oxford Living dictionaries 2020)

A common theme from these definitions is the emphasis on "human-like" characteristics and behaviors requiring a certain degree of autonomy such as learning, understanding, sensing, and acting. They, however, do not provide a framework to underpin such behaviors. This is problematic because humans, whose behavior AI systems or agents are supposed to mimic or in certain cases act on behalf of, behave according to a number of principles and standards such as social norms. Social norms can be argued to provide a framework to navigate among all the behaviors that are possible in any given situation. They introduce the notion of acceptable behavior, because they determine the behaviors that "others (as a group, as a community, as a society ...) think are the correct ones for one reason or another" (Saadi 2018). As a result, socially accepted behavior is central to how we act in a given context or environment. This suggests that the definitions of AI presented above are somewhat incomplete, because the AI agent or system has no way of determining which behavior is acceptable among those that are possible without such an equivalent framework.

While fictional, Asimov's three laws of robotics probably represent one of the first attempts to provide artificially intelligent systems or agents with such a framework. Attempts to create new guidelines for robots' behaviors generally follow similar principles.[4] However, numerous arguments suggest that Asimov's laws are inadequate. This can be attributed to the complexity involved in the translation of explicitly formulated robot guidelines into a format the robots understand. In addition, explicitly formulated principles, while allowing the development of safe and compliant AI agents, may be perceived as unacceptable depending on the environment in which they operate. Consequently, a comprehensive definition of AI must also provide a flexible framework that allows AI agents or systems to operate within the accepted boundaries of the

---

[3]www.pwc.in/assets/pdfs/consulting/digital-enablement-advisory1/pwc-botme-booklet.pdf

[4]https://epsrc.ukri.org/research/ourportfolio/themes/engineering/activities/principlesofrobotics/

community, group, or society in which they operate. By doing this, activities of AI agents or systems designed under such a framework naturally allow other stakeholders to regulate their activities, too. As a consequence, we choose to adopt a new, extended definition in this book: by AI, we understand any system (such as software and/or hardware) that, given an objective and some context in the form of data, performs a range of operations to provide the best course of action(s) to achieve that objective while simultaneously maintaining certain human/business values and principles.

AI (or more generally data science (DS)[5]) holds the potential to provide new and often better approaches for solving complex problems in almost every aspect of everyday life. While there is no single definition of AI, in this book, it is defined as stated above.[6]

Note For the sake of simplicity of this book, the terms DS and AI are used interchangeably. Similarly, AI algorithm and machine learning (ML) algorithm are used interchangeably.

AI is defined as any system (such as software and/or hardware) that, given an objective and some context in the form of data, performs a range of operations to provide the best course of action(s) to achieve that objective while simultaneously maintaining certain human/business values and principles.

Organizations often rely upon AI to process and find patterns in large volumes of data, which can in turn lead to innovation, new insights, and improvements in organizational performance and can also help firms create a competitive advantage over market rivals. As of today, there are countless examples of where AI is already being used for such purposes from a variety of industries. For example, law firms specialized in litigation might use AI to process and review large number of contracts, which speeds up their contract review process and helps them deliver more cost-efficient services to their clients.[7]

Another example can be observed in the retail industry, where ecommerce organizations use AI to boost their sales through recommendations. To illustrate, the ecommerce leader Amazon uses AI to power its recommendation

---

[5]While one may argue the differences between AI and DS, from a practical standpoint AI is an integral part of the DS process.

[6]Systems (such as software and/or hardware) that, given an objective and some context in the form of data, perform a range of operations to provide the best course of action(s) to take to achieve that objective while simultaneously maintaining certain human/ business values and principles.

[7]For example, https://digital.hbs.edu/platform-rctom/submission/jp-morgan-coin-a-banks-side-project-spells-disruption-for-the-legal-industry/

engine, which is associated with 35% of all purchases on the Amazon website.[8] Banks and insurance companies rely on virtual assistants to increase their productivity and create new and cost-effective ways to serve and interact with their customers. Similarly, in the healthcare industry, hospitals and other health organizations may use AI to automate their workflow and reduce the number of unnecessary hospital visits.[9] Likewise, pharmaceutical companies may use pattern recognition to create new drugs and develop sophisticated image processing techniques to help improve doctors' diagnostic capabilities and reduce the amount of time required for accurate diagnosis of expensive-to-treat, life-threatening diseases (Ganesan et al. 2010; Alizadeh et al. 2016; Leonardo et al. 1997; Lyons et al. 2016).

Looking at these examples, it is not surprising that AI is gaining traction in nearly every industry, including, but not limited to, manufacturing, retail, healthcare, life sciences, and legal. The future holds many exciting possibilities in these fields, and research is putting once futuristic ideas into practice. Many would consider the fact that self-driving cars are now being trialled out on our roads to illustrate this point.

This acceleration of the use of AI is amplified by the widespread availability of AI technology vendors, with many offering cheap and easy-to-integrate state-of-the-art AI tools into existing products and solutions, and the research and development of new business processes and solutions.

However, the use of AI in some, if not all, of these industries presents multiple challenges, in terms of governance and in relation to the safety and the liability of devices and system equipped with it, that need to be addressed. To illustrate, AI solutions in the healthcare industry, along with supporting technologies such as Internet of Things (IoT) devices, represent the biggest market opportunity in the foreseeable future, according to Allied Market Research.[10] Yet, it is likely going to present one of the most challenging environments in relation to fundamental rights, patients' safety, and efficacy of devices used (Char, Shah, and Magnus 2018): AI-powered healthcare products inherently rely on sensitive patient records, giving rise to data privacy concerns. Additionally, such data is likely to be population specific, meaning that AI solutions may scale poorly.

---

[8]www.mckinsey.com/industries/retail/our-insights/how-retailers-can-keep-up-with-consumers

[9]builtin.com/artificial-intelligence/examples-ai-in-industry

[10]According to Allied Research, the US healthcare IT market is projected to reach $149.17 billion by 2025 at 11.7% CAGR.
www.prnewswire.com/news-releases/u-s-healthcare-it-market-to-reach-149-17-billion-by-2025-at-11-7-cagr-says-allied-market-research-839141421.html (Accessed on 2019-10-02)

Challenge of Artificial Intelligence

The potential impact and transformative potential of AI are undeniable. This technology, like many other pioneering technologies, may however incur undesired side effects, and perhaps more often than expected. One such side effect may be the perpetuation of biases in society. Examples include, but are not limited to, Google's auto-labeling image recognition algorithm, labeling black people as "gorilla,"[11] and Amazon's recruitment AI system being gender biased.[12] Additionally, AI-powered solutions could be entrusted to make decisions whose gravity it is unable to understand. For example, in the healthcare sector, an AI-powered solution giving an incorrect diagnosis could become responsible for life-endangering outcomes. This was raised in the context of IBM's Watson Health which gave wrong recommendations on cancer treatments that could cause severe and even fatal consequences.[13]

It is therefore necessary to develop a business and regulatory environment that ensures

- That organizations at the forefront of AI research and development can maintain their competitive advantage.

- That organizations, at the same time, protect citizens and the environment. In other words, the challenge is to create an environment that encourages and nurtures innovation rather than impede it while protecting citizens.

These two points will be considered to define "sustainable AI" in this book.

Efforts to move in this direction are being made in western countries, for example, with the introduction of the General Data Protection Regulation (GDPR) in Europe and equivalent/similar regulations in other developed countries. While this is a step forward, GDPR is just the beginning of a long list of constraints and rules that modern AI organizations will be subjected to in the near future. Developing countries however, which present a unique opportunity for the development and deployment of AI-powered products, are yet to follow suit. One such opportunity in developing countries is in the

---

[11]https://eu.usatoday.com/story/tech/2015/07/01/google-apologizes-after-photos-identify-black-people-as-gorillas/29567465/
[12]www.telegraph.co.uk/technology/2018/10/10/amazon-scraps-sexist-ai-recruiting-tool-showed-bias-against/
[13]www.statnews.com/2018/07/25/ibm-watson-recommended-unsafe-incorrect-treatments/

healthcare sector, where the severe health workforce shortage continues to stress the already inefficient and often too expensive public health system. AI and telemedicine could help tackle some diseases and reduce stress on the public health system.[14]

The implementation of GDPR provided some companies with an opportunity to conduct extensive audit of their data ecosystem, what would have been a step toward addressing some of the challenges highlighted earlier; however, on the grand scheme of things, most organizations, especially small and medium-sized enterprises (SMEs) and businesses (SMBs), and traditionally nontechnological organizations either struggle to or are yet to define, develop, and/or implement processes along with good practices leading to a sustainable, safe, and ethical use of artificial intelligence.

Sustainable Artificial Intelligence

To better understand how AI challenges discussed thus far can be addressed through a sustainable practice of AI, it is essential to formally define the concept of sustainable AI or DS. There is no single definition to the concept of sustainable AI. One can think of sustainable AI/DS as AI subjected to organizing principles, including, but not limited to, processes which could be organization specific, regulations, best practices, and definitions/standards for meeting the transformative potential of DS while simultaneously protecting the environment, enabling economic growth, and social equity.

Note that the notion of sustainable AI is inherent to the definition of AI provided in "The Need for Artificial Intelligence" section.

From the above definition, it is clear that organizations committed to the sustainable development and deployment of AI systems will have to comply with some rules and be subjected to a certain number of constraints. Some of the abovementioned constraints might be application specific, which means that organizations have to understand what is relevant to their businesses. This can be more expensive and difficult for some businesses depending on their level of maturity, understanding of elements that enable AI and industry in which they operate. For example, an AI system that predicts the energy consumption of a user probably won't follow the same design/development constraints as an AI system that assists in the diagnosis of cancer or powers a self-driving vehicle.

---

[14]www.dynamicsfocus.com/8441/forus-health-uses-ai-to-help-eradicate-preventable-blindness/

This book introduces the sustainable artificial intelligence framework (SAIF), a framework to help organizations

- Design and develop sustainable AI systems and/or improve their understanding of elements that enable AI

- Size the impact of AI/DS on their ability to create value in the short, medium, and long term

- Anticipate future regulations or policies to ensure that they do not impede their competitiveness and ability to innovate

- Audit their current AI systems

This is accomplished through four pillars, through which the social, economic, and political implications of AI systems are integrated as inherent aspects of the design and deployment of AI systems. These pillars consist of the human factor, a common intra-organizational understanding of AI, AI system governance, and performance measurement, further discussed below:

- **Human factor**

  The human factor pillar aims to provide a framework for understanding and assessing to what extent an AI system affects its users and develops methodologies or tools to protect users of such systems.

- **Intra-organizational understanding of AI: toward transparency**

  A thorough conceptual understanding of AI by business decision makers is a prerequisite for the development of an environment that nurtures DS innovation rather than impede it. Similarly, data scientists need to develop a better understanding of the principles and business implications of the AI system they are developing. The intra-organizational understanding of AI focuses on developing methodologies to help business decision makers and data scientists within an organization gain a better common understanding of their AI systems.

- **Governance**

  Understood as the governance of AI systems within an organization, AI system governance helps to establish accountability, responsibility, and oversight while ensuring that individuals responsible for AI systems have the right expertise for the task at hand.

- **Performance measurement**

  Performance measurement is concerned with the definition of clear and transparent performance metrics of the AI system which reflect the values of the business.

Conclusion

Due to its recent success, AI is gradually becoming the cornerstone of any modern organization's operational strategy. This is justified by a significant increase in the adoption of the AI technology over the past recent years. However, despite its potential, the AI technology introduces several challenges that organizations adopting the technology must address to reap its benefits. The concept of sustainable AI provides elements through which an organization can effectively address such challenges and manage the risk around AI.

Ethics of the Data Science Practice

Ethics in Data Science

Chaos isn't a pit. Chaos is a ladder. Many who try to climb it fail and never get to try again. The fall breaks them. And some are given a chance to climb. They refuse, they cling to the realm or the gods or love. Illusions. Only the ladder is real. The climb is all there is.

—Petyr Baelish, *Game of Thrones*

The above quote describes chaos as an instrument for accessing and maintaining power. A strategy that is employed by various protagonists for their own gain throughout the HBO drama *Game of Thrones*. *Game of Thrones* presents an attention-grasping depiction of what human civilization looked like centuries ago, before the introduction of laws and rules deterring people from acting in a manner that negatively affects others. The drama highlights how

© Ghislain Landry Tsafack Chetsa 2021
G. L. Tsafack Chetsa, *Towards Sustainable Artificial Intelligence*,
https://doi.org/10.1007/978-1-4842-7214-5_2

people with power use this to their advantage. Today's society is far different from that of *Game of Thrones*. Importantly, one no longer gets away with just any behavior. Instead, there are rules governing one's actions and behavior.

While *Game of Thrones* is fictional, some comparisons can be drawn to the evolution of AI. More precisely, with the advent of social media, the proliferation of Internet of Things (IoT) devices, and digitalization, access to data has never been easier. One can access data about people's sleep pattern; what they do first thing in the morning; how fast they drive, walk, or type on a keyboard; etc. Moreover, data processing software and hardware have significantly improved. This results in a drastic increase for possibilities for the analysis and interpretation of this data. These two factors combined suggest that, beside one's imagination, there is virtually no limit to what one can do using AI. However, this does not mean that all these opportunities to exploit data are aligned with a society's values. In the absence of laws and regulations to help navigate the possibilities, how does one decide what to do and not to do? Fortunately, the study of ethics can be argued to provide frameworks to guide one's actions.

This chapter explores ethics of AI and is organized as follows:

- First, we examine the relevance of ethics in AI.
- We then discuss AI inferencing capability and its ethical implications.
- Next, we have a discussion on data as a business asset.
- Finally, we make some concluding remarks.

Ethics and Their Relevance to AI

The Oxford English Dictionary defines ethics as[1] "moral principles that define a person's behavior or the conducting of an activity." In line with this, ethical decision-making involves adhering to a "code of conduct" for guiding one's activities which requires consideration of others' values, beliefs, and consequences.

On the one hand, from a societal perspective, it is expected that certain ethical behaviors are respected by citizens regardless of individual characteristics. Such characteristics may include religion, geographic location, and education. Ethical behaviors expected in society are often referred to as societal ethical behavior and include fundamental governing rules such as respect for another's property, refraining from violence against another, and treating others with civility.[2] Many such rules are captured by the legal

---

[1] www.lexico.com/definition/ethics
[2] https://examples.yourdictionary.com/code-of-ethics-examples.html

framework of a society. However, rules provided by the law cannot tell us what to do in relation to every dilemma we may face and every decision we may take. As a result, not all unethical behavior is illegal. However as argued by Whittingham (Whittingham 2008), when left unchecked, unethical behavior can easily transcend into illegal behavior. On the other hand, from an organization's perspective, one can think of ethics as a set of rules and/or principles governing how organizations conduct (business) activities.

In the professional context, ethical behaviors that can be expected by society are often described and captured in agreements with relevant decision makers. Such rules are defined by the organization, but are also influenced by the industry in which an organization operates, and regulations which prescribe whether certain behaviors are appropriate or acceptable when dealing with customers, colleagues, and other organizations. Examples include the ethics of medical practice, the code of judicial conduct, and companies' employee handbooks. We will refer to organizational and societal ethics as the "general code of conduct."

The general code of conduct is generally well established and understood. However, technological innovation through research and development (R&D), especially when concerned with exploring the potential of cutting-edge technologies, may test its boundaries. This is illustrated through the concept of privacy defined as

> *The ability of an individual or group to seclude themselves, or information about themselves, and thereby express themselves selectively.*[3]

Privacy is constantly challenged by the ubiquity of data collection and processing opportunities and the dramatic expansion of the ability to carry out online inquiries about individuals. This complex relationship between privacy and cutting-edge technologies is summarized by Andrew Grove, cofounder and former CEO of Intel Corporation, as

> *[...] one of the biggest problems in this new electronic age. At the heart of the Internet culture is a force that wants to find out everything about you. And once it has found out everything about you and two hundred million others, that's a very valuable asset, and people will be tempted to trade and do commerce with that asset. This wasn't the information that people were thinking of when they called this the information age.*[4]

It is, therefore, not surprising that AI systems have been at the center of criticism on ethical grounds as we continue to understand and witness their outcomes. To better understand the basis of such concerns, it is essential to explore further the key elements of AI systems.

---

[3]https://en.wikipedia.org/wiki/Privacy
[4]www.esquire.com/entertainment/interviews/a1449/learned-andy-grove-0500/

AI systems are generally built around two basic elements: a knowledge base and an inference capability. Through inference, an AI system uses knowledge gathered during its training to infer things (generally new knowledge) about new data. Typically, a knowledge base is made up of a wide array of information or data (facts, relationships, theories, etc.) relevant to certain aspects of the world. Computer programs are written to provide computers with the ability to manipulate and navigate the knowledge base to identify or discover patterns. These patterns are then used to suggest a course of action to achieve a given objective. Specifically, this process, resulting in a trained algorithm or model, is referred to as "learning" or training. As a result, computer programs that perform such tasks are known as "learning algorithms" or more formally "machine learning algorithms." As in any matter involving learning, increasing AI systems' knowledge base generally leads to better results. Consequently, real-life AI systems tend to require large volumes of data for optimal functioning. However, this is only true where data quality is "good."[5] Conversely, poor-quality data ultimately leads to poor AI systems. This is commonly referred to as "garbage in garbage out" within the AI community. In other words, an AI system is only as good as the knowledge base that was used to create it.

From what precedes, it is essential when building an AI system to rely on data that, as best as possible, accurately represent the real-world problem the AI system is trying to model. Achieving this often involves using sensitive information, which in turn raises a number of privacy concerns around the acquisition, handling, and processing of such information. This has been widely acknowledged by various institutions via the general code of conduct, which in some countries[6] has evolved to regulate the information flow between users and various organizations or service providers through regulation-like measures such as the GDPR.

One of the major restrictions (arguably one of the most important from a user's perspective) imposed by the GDPR is that, similar to a surgeon seeking consent from a patient when carrying out a surgical procedure, organizations under the GDPR must seek explicit consent before collecting and using user information. Yet a key difference between consent requested by the surgeon and an organization remains that when the patient consents, there is a duty to very clearly engage with the patient to understand what this consent implies. In other words, the surgeon is obliged to act in the patient's best interest, whereas a company is generally not obliged to do so.

As per GDPR, users can "exercise control" over their data by choosing to consent or not to. This control does not exist in practice because users often cannot use the service unless they agree to all terms and conditions in full.

---

[5]The assessment of data quality is inherent to the application domain at hand.
[6]This is problematic in most developing countries, which are yet to follow suit.

Additionally, once users have given consent, they have little to no control over, let alone any real understanding of, what the data is used for and how it is used. This is arguably problematic, given that organizations generally do not invest much in ensuring that users understand what consent implies. This becomes clear when looking at terms and conditions, which now include statements in the form of "your data will only be shared with trusted partners"; who those partners are, what they do, and the nature of the agreement between them and the organization seeking consent are generally a mystery to the user giving consent.

For example, it has been reported that some applications collect users' personal data which are then shared with Facebook for advertisement targeting.[7] Moreover, by providing a login service via Facebook, Cambridge Analytica – which was at the center of the Facebook-Cambridge Analytica data scandal – was able to collect sensitive information about millions of Facebook users and associated friends, which it then used to profile them for political campaigning. A practice deemed unethical not only because it can be seen to violate users' rights to privacy but also because of the misuse of users' trust. The concepts of trust and ethics are inextricably linked, as illustrated by Hosmer's definition of trust which makes specific reference to the importance of ethical behavior:

> Trust is the expectation by one person, group, or firm of ethically justifiable behavior – that is, morally correct decisions and actions based upon ethical principles of analysis – on the part of the other person, group, or firm in a joint endeavour or economic exchange. (Hosmer 1995)

It is clear that an organization's image or business may be damaged when users perceive ethical standards to be low. Following the Facebook-Cambridge Analytica data scandal, Facebook saw its market capitalization go down by more than $100 billion, while Cambridge Analytica ceased operations. However, where the true ethical standards of a company remain opaque, customers may be unaware that their privacy is at risk.

The above discussion illustrates that it is fundamental to understand what it means to consider ethics in the context of data science. The following sections discuss ethical requirements for the practice of data science.

---

[7]www.wsj.com/articles/you-give-apps-sensitive-personal-information-then-they-tell-facebook-11550851636

Ethical Nature of AI Inferencing Capability

The use of AI's inferencing capability does not necessarily have ethical consequences, since the inferred information may be ethically neutral. In other words, the analysis of information may only have ethically unrelated effects. For example, an AI system that uses past weather conditions to infer weather conditions in the future may do so without any ethical consequence. Conversely, an AI system that labels dark-skinned individuals as "gorillas" does have ethical consequences because of the biased assimilation of people with gorillas.

Another example of AI inference with ethical connotations is provided by the debate around the use of criminal risk assessment tools. These tools take as input an individual's details and output a "recidivism" score indicating the likelihood of that individual reoffending. A judge then uses the score along with other factors to decide on the fate of the affected individual. While a high score suggests a severe punishment, a low score calls for the leniency of the judge and ultimately suggests a less severe punishment. Criminal risk assessment systems rely on historical crime data to build their inferencing capability. Therefore, communities (especially minorities) that have historically been targeted by law enforcement are at an increased risk of scoring high on the risk assessment scale than others. As a result, risk assessment systems are likely to perpetrate the already existing bias toward those minorities. Stated differently, what the risk assessment system suggests is that people be punished more harshly because they might reoffend. A reasoning that one can argue is morally flawed in essence because punishment is for past behavior rather than a behavior that may or may never materialize.

Similarly, an AI system that infers personal features such as sexual orientation, race, religion, and political preferences has ethical consequences either because of the privacy-related nature of the disclosed information or how it may be used. Irrespective of the level of accuracy of the prediction, disclosing such information could be discriminating in some circumstances.

Beside privacy-related ethical consequences of AI systems, we can roughly attribute ethical consequences of AI systems to the following factors:

- Characteristics that influence the inferred information are for the most part derived from a limited number of observations. This reflects a problem known as the "law of small numbers" in which the resemblance between a small sample and the larger population from which it is drawn is often exaggerated (Rabin 2002). For example, in a fair coin toss experiment, one may observe 70 heads and 30 tails, after tossing the coin 100 times, and conclude that a fair coin toss has 70% of landing heads and 30% of landing tails, which is incorrect.

- The knowledge base is biased to the advantage of a specific group. Risk assessment systems discussed earlier represent a good example of scenarios where the knowledge base is biased to the advantage of a specific group. This is fairly common because the knowledge base is often just a digital representation of our own perception of the world.

- The knowledge base may contain errors. Companies often rely on third-party data for making critical decisions. For example, creditors, insurers, employers, and other businesses rely on a credit score to evaluate applications for credit, insurance, employment, or renting a home. This is problematic given that credit data is collected from multiple sources and often contains errors. As a result, affected individuals may appear riskier than they are, which in turn means that they will be offered a higher interest rate. Additionally, because affected individuals are often unaware of what data is being used, opportunities for improvements are limited.

- The knowledge base is made up for the most part of interpersonal subjective data. This type of information, at best, only provides a "crude" orientation and is not comparable across individuals. Such information includes recorded perceptions. For example, when asked to rate their pain on a scale from 1 to 10 (1 being the lowest level of pain and 10 the highest), a patient may deliberately give a number that increases their chances of getting a painkiller. Irrespective of what score is given, it only makes sense to the patient rating the pain.

- Lack of transparency regarding the relevance of the inferred information. Many learning algorithms are probabilistic;[8] however, because of the existing gap between data scientists and the business, set probabilities are often misinterpreted or (in some way or form) converted into grades or scores. Such scores are then used to influence how a service is delivered to its users. For example, a criminal risk assessment system may compute a recidivism score which is then translated to a

---

[8]In simple terms, one can think of a probabilistic algorithm as one for which the operating criteria are not clearly defined by the programmer.

risk level associated with an individual. Unfortunately, because of the proprietary nature of most systems powered by learning algorithms, it is impossible to rule out flaws in their design.

- Data processing errors. Errors in data processing may occur for many reasons ranging from input to data manipulation errors. For example, an organization would like to understand how it is perceived by its customers by analyzing social media posts (short paragraphs of texts), some of which may include sarcasm. However, while significant progress has been made, techniques used to perform such tasks are not perfect, so errors are expected and one should be aware of that.

- Misleading or false advertising of AI products. For example (at the time of the writing of this book), in the healthcare industry, there is currently a lot of excitement about AI diagnosis tools, all of which at best can only serve as assistants to doctors. Unfortunately, they are rarely publicized as such and may instead be understood by users to fully replace the services of a medical professional. This could cause serious damage, especially in developing countries where regulations and controls are loose or nonexistent and where such services may be in high demand as a result of the widespread lack of (access to) qualified medical staff.

- Changes in data. Changes in the data population render the knowledge base obsolete, meaning that it becomes a poor representation of the new target population. Organizations have to thus enable regular audits on the population data to ensure correct representation.

Data – The Business Asset

Digital transformation – transforming business processes to integrate the use of new digital technologies such as cloud, mobile, big data, and social networks – is steadily becoming the de facto operational strategy for organizations looking to optimize existing operations while capitalizing on new opportunities.

As a result, the amount of data generated by users through their interactions with digital services or platforms (for simplicity, we refer to it as user-generated data) has skyrocketed over the past few years. It is estimated to reach 175 zettabytes in 2025 according to market intelligence company IDC.[9] Meanwhile, the number of Internet users worldwide jumped from 2.4 billion to 4.4 billion from 2014 to 2019,[10] allowing an even greater number of individuals to access Internet-based services.

Despite this, most organizations are still trying to figure out how to create value from their share of user-generated data. Technology pioneers, especially Google, Amazon, and Facebook,[11] have been the main beneficiaries of the increasing availability of "big data." To illustrate, Amazon recorded historically high advertising revenues of $3.6 billion in Q3 2019. The ecommerce giant achieves such performance by ensuring the ads it portrays on its platform on behalf of companies are "relevant" to the users they are delivered to. Yet, the key aim of determining this "relevance" is to maximize profit. A prerequisite for this is having enough meaningful information, such as on past searches and purchases, about such users.

Other examples involve companies using "freemium" services, which are offered to the customer free of charge at the point of use but provide a company with data that can later be monetized, for example, by offering the customer paid services. In the case of Google's free search engine, Google monetizes its information on searches by selling advertising space to other companies.[12] It is, therefore, not surprising that organizations, rightfully, now perceive data as a business asset. This results in a trend as part of which individuals (or at least the data they generate) can be thought of as being assimilated into commodities. One could argue that organizations actively devise schemes for collecting and monetizing user-generated data. Such schemes are often opaque, given that the service as presented to and understood by its users often does not provide full disclosure of what it is really doing.

Companies may argue that users benefit from providing companies with access to their data, as it allows the company to provide a better or more targeted user experience. However, it is often questionable whether the value of the improved service received is proportional to the value of the data provided to companies. An extreme example is provided by services that require users to register/sign up, but where the content/service they receive

[9]IDC White Paper – #US44413318, November 2018
[10]www.internetworldstats.com/stats.htm
[11]Alongside Apple and Microsoft, they are often referred to as GAFAM
[12]https://marketingland.com/youtube-kicked-in-15-billion-as-google-ad-revenues-topped-134-billion-in-2019-275373

is virtually identical whether they register or not. The industry is well aware of the misleading nature of such practice, and some companies are beginning to respond to calls for greater protection. To illustrate, Apple actively monitors and rejects applications exhibiting such behavior from its App Store. As of the 26th of October 2020, its terms and conditions state

> [...] if your app doesn't include significant account-based features, let people use it without a log-in. Apps may not require users to enter personal information to function, except when directly relevant to the core functionality of the app or required by law [...].[13]

Organizations monetizing user-generated data often only present a skewed reality to their user, which can be perceived as a lack of transparency, in the sense that by joining together data from various users, they filter what is shown to its users and ultimately shape what each individual sees, thinks, and does. For example, consider the case of search engines: People increasingly rely on search engines when seeking for information and advice, in relation to almost all aspects of everyday life (Carroll 2014). Results from the search engine ultimately change our opinions and perception of the world. The apparent choices users are given are often misleading because what they ultimately choose from is at the organization's discretion (Goldman 2008; Jordan 2017). This represents a practice that most users, if given transparency, would likely disapprove of. In summary, there are several reasons why the way in which companies exploit the business asset that data has become is often conflicting with the values of the data subjects – the individuals from whom the data is collected.

Additionally, the ability to treat data as a business asset acts as a deterrent for operational transparency, specifically data handling and processing. Data subjects, if made aware of the real value of their data, could demand a share of the income generated from their data, thereby threatening business's profits. One could argue that most of the services the user received through which it provided its data are free of charge and can be seen as compensation for their data. However, despite efforts by some organizations to compensate users for their data, the current income distribution seems rather skewed toward the data handler.

As with any business asset, the governance of data is critical for ensuring that data is consistent across the organization, trustworthy, and is not misused or manipulated in a way that can harm either the organization handling the data or data subjects. In other words, data governance encourages good behavior and proactively limits behaviors that create risks through regular audits.

---

[13]https://developer.apple.com/app-store/review/guidelines/legal

These risks may be in relation to regulatory fine, data security, reputation damage, etc. For example, leaked sensitive information may place affected individuals under great distress including threats of blackmail, with horrendous consequences as was exposed with the data breach around the Ashley Madison extramarital affair site, following which suicides were reported.[14] Furthermore, implementing a proper data governance and audit strategy helps an organization answer questions such as: What happened to the data? Who accessed it? How is it being stored? How long has the data been within the organization? etc. Answering some of these questions may help organizations understand how well they comply, for example, with the GDPR Article (5) "personal data shall be kept for no longer than is necessary for the purposes for which it is being processed."

A key limitation of this article is that it is probably too vague and difficult to enforce by current regulators. As a result, an organization may be tempted not to adopt a data governance strategy; however, doing so can only fuel the trust crisis that exists in this day and age between organizations and customers. According to the 2018 second edition of the Salesforce State of the Connected Customer report, 59% of the customers and business buyers (out of 6700+ surveyed) think their personal information is vulnerable to a security breach, while 62% feel uncomfortable with how companies use their personal/business information.[15]

Given new regulations, data can easily turn into a liability if up-to-date standards around data governance – including security as an integral part – are not met. Incidents involving organizations such as British Airways,[16] Equifax,[17] and Marriott[18] are illustrative examples of data becoming a liability for the organization collecting it.

---

[14]www.theguardian.com/technology/2016/feb/28/what-happened-after-ashley-madison-was-hacked

[15]www.salesforce.com/content/dam/web/en_us/www/documents/e-books/state-of-the-connected-customer-report-second-edition2018.pdf

[16]https://ico.org.uk/about-the-ico/news-and-events/news-and-blogs/2019/07/ico-announces-intention-to-fine-british-airways/

[17]www.equifaxbreachsettlement.com

[18]https://ico.org.uk/about-the-ico/news-and-events/news-and-blogs/2019/07/statement-intention-to-fine-marriott-international-inc-more-than-99-million-under-gdpr-for-data-breach/

Conclusion

The past few years have witnessed the ever-increasing development and deployment of AI-powered applications in almost every aspect of the society from public and private institutions to government agencies. This generally provides efficient approaches for solving complex problems; unfortunately, the way the technology is developed and deployed may lead to unwanted and undesired consequences in relation to privacy, fairness, trust, and transparency. Fortunately, the discussion of ethics provides a framework to help understand these consequences. More generally, ethics helps unveil aspects of the AI development and deployment process that present opportunities for unwanted behaviors. Unveiling these ethical hazards requires an understanding of the fundamental elements of AI along with their implications.

The following chapters introduce methodologies for minimizing ethical consequences of AI highlighted in this chapter.

Sustainable AI Framework (SAIF)

Overview of SAIF Framework

The next few chapters discuss components of the sustainable artificial intelligence framework (SAIF). As highlighted in Chapter 1, SAIF relies on four pillars: the human factor, the intra-organizational understanding of AI, governance, and performance measurement. As seen in Figure 3-1, these four pillars form the conceptual founding block the operating model of the SAIF framework is built on. Through this operating model, an organization can dictate the ultimate behavior of an AI system. Specifically, the SAIF operating model aims to allow an organization to

G. L. Tsafack Chetsa, *Towards Sustainable Artificial Intelligence*,
https://doi.org/10.1007/978-1-4842-7214-5_3

- Understand the current and (potential) future risk profile and exposure to unwanted consequences by assessing the impact of AI on the organization's ability to create value in the long, medium, and short term

- Improve the intra-organizational understanding of elements that enable AI within the organization

- Adopt a practical approach for the design, development, and deployment of AI systems that incorporate both business and human values and principles

- Audit an organization's existing and future AI systems

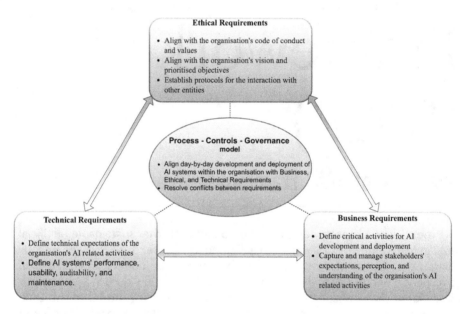

Figure 3-1. SAIF operating model

At the core of SAIF's operating model are three fundamental elements: the AI development process, controls, and governance. They ensure that an organization's vision is aligned with the day-to-day development and deployment of AI systems within the organization.

- **AI development process:** This consists of steps to be taken in the development and deployment of AI systems.

- **Controls:** These impose a number of constraints on each process step which must be respected, helping to manage the end outcome and that it arrives at a solution that meets expectations.

- **Governance:** This defines the arrangement and responsibilities of different stakeholders through which the development process and controls are implemented within the organization.

The **process – controls – governance** model forms the core element of the SAIF operating model and may thus also be referred to as the SAIF core. It is discussed in greater detail in Chapter 4. These core elements are influenced by ethical, technical, and business requirements. Ethical, technical, and business requirements undoubtedly influence each other, and care needs to be taken to ensure all three factors deliver requirements that are consistent with each other.

Intra-organizational Understanding of AI: Toward Transparency

Mainstream public perception of AI varies greatly depending on an individual's perspective and experiences. On the one hand, from a user perspective, it can be perceived as a set of services that rely on data to enable new levels of innovations, insights, and organizational performance. On the other hand, from a more technical perspective, it can be perceived as a technology using mathematical frameworks, computing infrastructures along with associated software, and processing tools for analyzing and/or extracting patterns in

G. L. Tsafack Chetsa, *Towards Sustainable Artificial Intelligence*,
https://doi.org/10.1007/978-1-4842-7214-5_4

large volumes of data. Yet, each of these perspectives differs from our definition in "The Need for Artificial Intelligence" section of Chapter 1.[1]

The development and deployment of AI systems involves a degree of complexity not only from a technical and design perspective but also from a user's perspective. This highlights the importance of managing users' expectation and understanding of the system's functionality. It is unrealistic to expect users of AI-powered services to understand the intricacies of the underlying technology on which the services they receive rely. However, one would expect the service provider, as an organization, to understand the importance of managing the user's expectations in the same way legal firms are expected to understand the full complexity of services they provide, yielding a range of potential outcomes, and keeping their clients well informed throughout the process. In the case of a law firm, the leadership team is more than often made up of trained lawyers with years of experiences, which have the technical expertise required to oversee the delivery of legal services to their clients.

AI continues to find use cases in a wide range of industries and organizations as these realize its ability to improve the way they operate and interact with their clients. Most of these organizations, especially small, medium, and traditional businesses,[2] are not as tech savvy as mainstream adopters of AI. As a consequence, these companies generally lack a thorough understanding of AI. This may result in a lack of a common understanding and transparency between a business's decision makers and its AI development team.

This chapter discusses the process – controls – governance model of the SAIF operating model introduced in Chapter 3. As highlighted, the proposed approach aims to help organizations size the impact of such technologies and potentially limit their exposure to unwanted consequences resulting from their use of AI systems.

The remainder of this chapter is organized as follows:

- We first present the data science development process inspired by and enhanced from the cross-industry standard process for data mining.

- We next examine proposed controls imposed on the development process to help organizations assess and improve their understanding of elements that enable AI within the organization.

---

[1]Systems (such as software and/or hardware) that perform a range of operations to the data to provide the best course of action(s) to take to achieve a set objective while simultaneously maintaining certain human/business values and principles.
[2]Most organizations matching this profile generally rely on AI services offered by big technology companies such as Amazon, Google, and Microsoft.

- Third, we discuss the governance strategy or arrangement through which the development process and controls are implemented and maintained.

- Finally, we make some concluding remarks.

Data Science Development Process

There is only a limited literature about the DS process, which stands in contrast with other digital technologies such as software development. This results from the fact that DS processes are generally organization specific. Nevertheless, the existing literature broadly agrees that a typical DS development process involves the following key phases: problem formulation, data collection and/or interpretation, model building, evaluation, and deployment. A graphical representation of the DS process diagram, adapted from the cross-industry standard process for data mining (Shearer 2000), is shown in Figure 4-1. A brief overview of each of these phases is presented below. The newly introduced "performance metrics" phase, which allows the business and the development team to clearly specify the expected behavior of the system under development, is discussed in greater detail and will be revisited in Chapter 5.

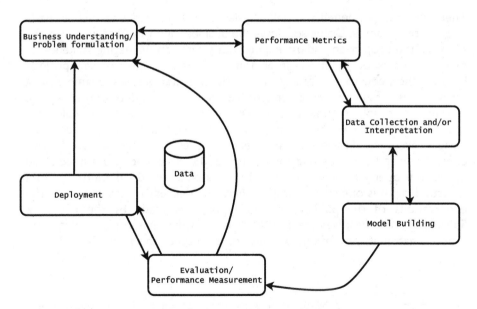

Figure 4-1. Data science development process

Problem Formulation

As its name suggests, the problem formulation phase aims to elaborate and define the boundaries of the problem at hand. The outcome of this phase is a research question or hypothesis that the AI team will investigate. To accomplish this, domain experts, business representatives, and the data science team collaborate to identify the business problem and translate it into a research question. This process may result in multiple research questions, which should be prioritized by the business leadership.

Performance Metrics

This phase of the development aims to ensure that the AI system has the right objectives. Although performance measurement is generally performed as part of evaluation, it often fails to integrate business and technical requirements not directly related to the AI system's ability to suggest a seemingly suitable outcome or course of action. For example, an AI recruitment system satisfying traditional performance metrics (e.g., accuracy, precision, recall, etc.) may fail to meet an organization's prioritized objectives, such as avoiding gender bias. Throughout the remainder of this book, we refer to an organization's prioritized objectives such as fairness as "soft performance metrics."

There are multiple benefits in defining performance metrics at such an early stage of the process. Doing so allows the business and the AI team to set the scope of the project and adjust expectations early on. Managing stakeholder expectations is of particular importance for the success of the project, because the need to satisfy any unrealistic expectations, for example, as a result of business pressures, often gives rise to the risk of deploying unreliable systems or simply setting the project up for failure. For example, the probabilistic nature of AI-powered systems means they rarely operate at perfect accuracy. However, users expect their applications to behave consistently and perfectly (Dzindolet et al. 2003), so expectations need to be managed properly to avoid and/or mitigate a potential disappointment. IBM Watson Health is one such example where the delivery meets neither of the expectations of the public and public promises made by the company. Specifically, IBM advertised that its IBM Watson AI platform would revolutionize cancer treatment, yet until now it has not lived up to this promise.[3, 4, 5]

---

[3]www.wsj.com/articles/ibm-bet-billions-that-watson-could-improve-cancer-treatment-it-hasnt-worked-1533961147
[4]https://spectrum.ieee.org/biomedical/diagnostics/how-ibm-watson-overpromised-and-underdelivered-on-ai-health-care
[5]www.theverge.com/2018/7/26/17619382/ibms-watson-cancer-ai-healthcare-science

In addition, the criteria that can be used to judge the operation and/or expected behavior of the system should be identified and documented as part of the performance metrics phase. As a result, the development team is provided with clear objectives. For example, in the case of a facial recognition system, one might consider, among other things, racial unbiasedness as a relevant soft performance metric. With that knowledge, the development team can then design a system that meets the organization's expectations starting with the data collection right until choosing which algorithms to use.

In some cases, the outcome of this phase might be that the project is deemed not to be feasible given the expectations. Using the example of an AI recruitment system, it may be unrealistic to expect to be able to obtain accurate data as to why a given employee succeeds or fails within an organization.[6] Algorithms may then simply reflect the current state of affairs, which may suffer from the biases one is trying to fix, given that they analyze historical information. As a result, it might be impossible to deliver a system that is unbiased.

Deciding which soft performance metrics reflecting business and ethical requirements to use requires ethical, economical, and often political reasoning, which is usually context specific and subject to individual preferences. As a result, another benefit of defining soft performance metrics up front lies in that one may choose to rely on algorithmic approaches to help maintain consistency across the system. For example, a person's understanding of fairness may vary from one individual to another. This means that systems designed by different individuals may satisfy different interpretations of fairness within the same organization. A consistent organization-wide definition or interpretation of fairness helps mitigate disparities, assuming it is possible to agree on such a definition. Ideally, such definitions should be established through an algorithm wherever possible. Doing so allows for a more consistent and sustainable definition of performance metrics.

Soft performance metrics should be extended to other important metrics related to the operation of the AI system and more generally requirements which are sometimes considered only as an afterthought.

These metrics may include compactness, scalability, dependability, and security, where

- **Compactness** describes a system that can be deployed with limited cost, energy, and bandwidth.

- **Scalability** is a system which can grow to manage increased demand.

---

[6]The mere identification of factors contributing to employee success or failure within an organization is still a big research question in the management literature (Rusu, Avasilcai, and Huţu 2016; Rafique et al. 2017).

- **Dependability** characterizes a system which is available as needed while maintaining a high degree of integrity. This is of particular importance for autonomous systems, such as self-driving cars that may rely on GPS signals to function properly.

- **Security** does not just relate to ensuring data privacy, but recognizes that security concerns should spread across all stages of the development process. For example, building scalable models often requires relying on a third party for computation, giving rise to risks related to third-party access. One should understand the security implications of each scenario, especially when dealing with issues that may have implications for human safety. For instance, some providers may keep a copy of information gathered and often allude to that in their terms and conditions. Necessary precautions should be taken to ensure that only authorized individuals, vetted to adhere to agreed rules, have access to such data.

Data Collection and/or Interpretation

At this stage of the process, the data needed to investigate the research question or hypothesis resulting from the initial phase is collected. Collecting data specific to the question at hand is traditionally achieved through an experiment. This may not, however, be the case with big data, where the data may already be available.

In this case, the data science team will need to work with business representatives and domain experts to understand what the data represents and its validity. As part of this so-called "interpretation phase," the data science team also goes through what is often referred to as the "data cleaning" process. This involves extracting relevant information and identifying key patterns from the data. In addition, the data is assessed against performance metrics identified earlier. A key benefit of defining performance metrics ahead of this stage of the process is that the development team can carry out the data collection effort with those metrics in mind. Using the fairness performance metric example from the previous section, a well-designed data collection strategy can help mitigate bias because skewed data is generally one of the main contributors to bias in relation to AI systems.

A common problem that one may experience at this stage is the incentive to reformulate hypotheses, defined earlier, based on data availability. This is usually not advised, as hypotheses derived from data may not generalize well. Also, there could be an incentive to reformulate the hypothesis as another hypothesis may be easier to prove, which risks the overall usefulness of the exercise. Using the example of the facial recognition software: Upon exploring data that an organization holds on some of its customers, it may conclude that it is possible to perform facial recognition and proceed forward to develop and deploy a facial recognition AI system for the general public. This can be problematic because of the potential differences between the data from which the hypothesis was generated and data that is collected on final users.

Model Building

The output of this phase is a candidate solution to the research question identified during the problem formulation phase. Proposed solutions are designed to best meet performance metrics jointly defined with the business. Similar to the data collection phase, a clear definition of performance metrics ahead of this phase allows the development team to prioritize models that best meet them. For example, energy suppliers in the UK are required by the Office of Gas and Electricity Markets (Ofgem) to be able to explain energy consumption estimates to their customers, giving rise to a clear performance metric. This is a compelling reason for organizations to prioritize explainable and interpretable modeling solutions when relying on AI to estimate energy consumption. More generally, performance metrics can guide some of the decision-making during model building. However, it is still paramount to make the development team interact with business leaders and other important stakeholders on a regular basis, for example, through review meetings. This allows the development team to review progress and confirm that the emerging solution still meets expectations. It also allows them to test if they have a clear understanding of the system under development. Most importantly, regular communication with business leaders ensures that they have full buy-in into the solution that is delivered.

Evaluation

At the evaluation stage, one should make sure that the model is answering the business question identified at the problem formulation stage. Further, during this stage one should assess and document how well it is performing against performance metrics defined earlier in the process. In summary, the objective of the evaluation phase is to assess the generalization capability of the model from the previous phase. That is to assess how well it performs on unseen data – data that was not used when building the model. In other words, when carried out correctly, this phase will provide insights into how well the model may perform in real life. Overall, evaluation is just as important as any phase of the DS process.

Deployment

The deployment part of the process is where the model, subject to being deemed successful in the previous phase, is either shared with the wider audience or packaged onto a production grade system for its distribution. Depending on the solution, this stage may also involve continuous monitoring of the final system to ensure that it continues to meet the organization's priorities.

Soft performance metrics related to operational aspects of the systems should be assessed against the deployed system. Including metrics such as security, scalability, compactness, and dependability, these metrics need to be identified and documented in the performance metrics phase.

AI Development Process Controls

While the process element provides guidelines for the steps to be taken in the development and deployment of AI systems, controls provide a template to the integration of business and technical requirements. They achieve this by imposing a number of constraints that the process must respect at each step, helping to manage the end outcome. Specific controls may also help to expose nontechnical collaborators to the elements that enable AI within the organization. This further helps individuals involved in the project to dictate the standard of the final system. In other words, one can specify how an AI system should behave by defining specific controls that help enforce and maintain the desired behavior. For example, a control that imposes checks for fairness at the deployment stage of the development process helps an organization understand how well it meets that aspect of its social responsibility.[7] Another function of controls is to provide an organization with an understanding of the risk and level of exposure of the system under development. They do so by reducing the probability that undesired outcomes occur or even limiting the range of possible outcomes to exclude (some) undesired outcomes altogether. Using the fairness example, an organization may be criticized for unfair treatment of certain user groups. This can lead to loss of opportunities, users' trust, and/or regulatory sanctions. If controls are able to effectively impose constraints on the system only yielding fair outcomes, this risk is eliminated entirely. As one would notice in the following, appropriate controls depend on the target objective and can themselves take the form of processes.

The set of requirements discussed as part of the SAIF operating model, consisting of ethical, business, and technical requirements, all play a role in

---

[7]Assuming fairness is one of the organization's corporate values and ethical requirements on the system.

defining appropriate controls. In some cases, these three types of requirements may be at odds with each other. For example, imposing a fairness control on the system may come at the expense of efficiency or effectiveness of the solution. This points to the importance of trading off ethical requirements, business requirements, and technical requirements ahead of the definition of relevant controls to ensure that they are working in harmony toward the same overarching objectives. An organization ensures that such a trade-off between technical, business, and ethical requirements is established and maintained through its governance arrangement; this is discussed in greater detail in the "Governance" section.

In this section, we discuss controls applicable to different phases of the DS development process. No controls need to be defined at the business understanding stage because at that stage of the development, all parties (business and development team) must agree on the problem to be solved and thus have an aligned understanding of the project to be undertaken.

Controls for Performance Metrics

As discussed earlier, defining suitable performance metrics requires a joint effort between the development team, the business, and domain experts. As a result, the business objective remains the main control for performance metrics. Agreed performance metrics can be thought of as constraints imposed on the overall AI system under development. However, specific controls need to be defined and implemented to ensure that performance metrics are defined and measured appropriately. As a result, controls for performance metrics involve defining an organization-wide strategy or procedure for measuring the performance of AI systems. This strategy may be industry and/or application specific.

Controls for Data Collection and/or Interpretation

Lack of data is probably one of the biggest challenges in developing innovative data products. Most organizations have to go through a data collection or acquisition process. Typical controls at this stage include, but are not limited to

- Defining and documenting an organization-wide data collection process and processing guidelines. This is illustrated by questions such as: Is there enough evidence that records being merged are from the same person? Are there missing data? If yes, what are the subsequent assumptions? This control should highlight the implications of the data collection procedure used by the organization. For example, self-reported, subjective data should only be perceived as useful input as opposed to being used to make important decisions. Examples of subjective data include information on patients rating how much pain they are experiencing. While this gives an idea of how much pain they are in, this information is only valid for the affected patient.

- Assessing compliance with regulations such as the GDPR where applicable.[8]

- Assessing and reporting data quality along with its provenance,[9] type (e.g., interpersonal subjective data), target population group, and whether data subjects (individuals on whom the data is collected) are aware of what it is being used for, especially when the data is provided by a third party.

- Defining an organization-wide data governance strategy. This is to ensure that critical data is identified and properly classified, and the right access controls are implemented.

Controls for Model Building

Typical controls for the model building process include

- Documenting assumptions and motivations for the selected model. Doing so ensures that the developed model is motivated and justified by its performance against the agreed performance metrics rather than merely being the development team's favorite algorithm. In the case of the "privacy" performance metric, the development team may decide to build a model around a privacy-preserving algorithm through federated learning. In a federated learning setting, the machine learning model is trained using decentralized data residing on users' devices. In other words, each device of a selected number of user devices trains a model and sends it to a central server. The central server aggregates all models into a single model and sends it back to users' devices for further training. While a promising approach, federated

---

[8]In our view, organizations should seek to comply with such regulations regardless of whether they are legally required to or not.

[9]Data provenance can be thought of as a record trail that accounts for the origin of a piece of data together with an explanation of how and why it got to the present place.

learning is relatively new and introduces a new set of challenges (Bonawitz et al. 2019) that need to be documented and mitigated by the development team.

- Testing compliance with performance metrics such as interpretability. This can be assessed by having the team in charge of the development explain the relationship between random inputs to and outputs from the model. It is worth noting that in some situations, designing an interpretable model may be of little importance. In such cases, the development team should not be pressured to rely on an interpretable model when this is not necessary.

- Identifying and documenting attributes that are likely to increase the risk of undesired ethical consequences. While application specific, such attributes may include, for example, gender, age group, and ethnicity.

- Ensuring that the model generalizes well, where generalization refers to the algorithm's ability to perform well on unseen data. In other words, one should be concerned with ensuring that the (input) data left aside for evaluation and/or testing of the model can be used to accurately assess the model's ability.

Controls for Evaluation

Controls for the evaluation phase ensure that the model's performance is properly measured and accurately reported. This involves defining and documenting company-wide procedures and guidelines for evaluating and reporting the AI system's performance. As part of this process, the development team shares, with the business, details on aspects such as how performance of the AI system is measured, what data is used and how it conforms with the target population, and statistics on partitions of the development data used for testing and training. The process of reporting evaluation metrics properly guarantees that all involved parties are aware of the generalization potential of the produced model before moving it to production.

Controls for Deployment

Controls for the deployment typically consist of checking the systems against certain performance metrics. In addition, processes must be put in place to ensure that only authorized individuals have access to data going through the deployed system. This is necessary because the data product is often designed in such a way that the data is stored once the system is live, giving rise to risks of data security breaches. Note that this is not necessarily true for privacy-preserving algorithms. The life cycle of a software system does not generally end when it is in production; instead the system enters a maintenance phase where it is periodically assessed to ensure that it does not become obsolete. The same applies to AI systems, especially given their strong dependence on historical data. Consequently, the development team should provide the business with a reasonable maintenance plan for the system. Although application specific, this will typically include monitoring changes in data, model performance, and any other otherwise undocumented variations across the whole system.

Variations may occur when differences exist between the data that was used to build the model and the data it encounters in production. A traditional approach for reducing the chances of this happening is to split the data into training and testing sets during model development. It is therefore fundamental that the target population presents the same characteristics as that seen from the data during model building. Furthermore, the distribution of the data may change over time due to changes within its target population or feedback from the deployment act. For example, spammers may change the structure or contents of spam emails as a result of the deployment of a new spam detector. Monitoring ensures that such changes are identified and dealt with accordingly.

While splitting the data into testing and training sets helps to control changes during model development, specialized architectures such as auto-encoders may provide a dynamic approach to tracking changes and assessing data integrity (Baldi 2011) for deployed models.

Controls discussed in this section are in no way exhaustive and simply aim to provide the reader with a clear understanding of what is expected from them. The next section examines governance arrangements needed for the effective implementation of controls and the development process.

Governance

Thus far, this book has discussed two of the three core elements of the SAIF operating model. While these are instrumental for an organization's successful AI strategy, their implementation requires a clear understanding or establishment of an organization's responsibilities toward its stakeholders, understood as those who can affect or be affected by an organization's actions. This is where AI governance comes into play. It steers an organization's AI strategy toward a vision and makes sure that day-to-day development is in line with the organization's prioritized objectives. In other words, it orchestrates and manages an organization's AI strategy in a manner that ensures that AI development maintains an organization's values and principles while ripping off the benefits of the technology.

At its core, AI governance is a form of corporate governance and should be inherent to the governance arrangement of any organization with a digital presence. However, because of the complex nature of AI systems, it is often unclear what AI governance means and what is expected from it. This section discusses the significance of governance to the sustainable development and deployment of AI systems.

- We first provide a practical definition of AI governance and highlights what is expected from it.

- We next discuss key actors of AI governance along with their responsibilities.

- Finally, we examine practical approaches for assessing AI governance arrangement within an organization.

Expectations of AI Governance

According to the online business dictionary,[10] governance is defined as the establishment of policies and continuous monitoring of their proper implementation by the members of the governing body of an organization. This includes the mechanisms required to enhance the prosperity and viability of the organization.

AI governance aligns with the above definition: it aims to define, implement, and monitor structures, processes, and procedures for the development and deployment of AI systems within an organization. It paves an organization's way to a prosperous future by continuously promoting an organization's values and making sure that day-to-day operations are lined up with the organization's goals. Consequently, an effective AI governance strategy should improve an organization's performance, which may include operational efficiency, financial performance, and social responsibility. Conversely, poor AI governance, with some examples having been highlighted earlier in this book, can put an organization at risk or allow an organization to lose sight of its purpose. SAIF's AI development process and controls discussed earlier provide a framework for developing and deploying AI systems that meet certain values and principles, including ethical values. These values are instilled within an organization through its leadership. Consequently, one can think of AI governance as the arrangement through which the AI development process and controls implement and maintain an organization's values and principles through its services.

Through its process – controls – governance model, SAIF provides the development team and decision makers with tools to ensure that the final system meets the organization's values as well as common social standards

---

[10]www.businessdictionary.com

where required. More precisely, it allows both business decision makers (may include regulatory and legal advisers if any within the organization) and the development team to collaboratively define appropriate controls for each step of the development process and oversee their implementation through the governance arrangement. Assessing and managing risks is inherent to the use of the SAIF framework. For example, tighter controls generally reduce risk and exposure when implemented correctly, whereas relaxing controls do the opposite. To illustrate, an AI system for making hiring decisions may need to be constrained through its controls in order to optimize productivity, ethics, and legality (Lipton 2016). However, an organization may choose to only focus on productivity, therefore exposing itself to potential consequences associated with not meeting ethical and legal expectations. This example further illustrates the need for a continuous collaborative effort between the development team and other business decision makers.

To illustrate how the process, controls, and governance elements relate to each other, let us consider the problem of estimating the electricity consumption of a household in the UK. The process component of the model defines the steps to be taken to build a model to solve the problem. One of the first steps in any model building process is the data collection stage. Controls provide constraints on each step of the process to ensure that it arrives at a solution that meets expectations. Suitable controls relating to the data collection stage could involve the inclusion of diversified household profiles and customer-specific information to personalize the final model's estimates. The role of governance is to oversee both the implementation and maintenance of each step in the process as well as the respective controls. In terms of data collection, the governance arrangement may delegate the responsibility of establishing and maintaining oversight over the data collection strategy to the business representative, while making the development team responsible for assessing and reporting on its quality.

People and Values

An organization's leadership plays a central role in its AI governance strategy. More precisely, the organization's leadership is responsible for maintaining and coordinating the ongoing evolution of the organization's AI governance arrangements. It achieves this by delegating core oversight tasks to its staff and ensuring that individuals have the right skill set or that appropriate training is put in place. This is essential because delegating core oversight establishes accountability and responsibilities. Additionally, without clear responsibilities, there is no guarantee that the organization's governance approach is in line with its values. This is of particular importance as an organization becomes more and more successful in using data strategically.

While delegating oversight is instrumental for a successful AI governance strategy, one of the biggest challenges is ensuring that an organization has the skill set needed to define and implement the appropriate governance strategy. To achieve this, an organization's leadership can rely on a carefully selected and diversified AI governance committee. The committee, which could be under the supervision of the data/AI director, is then tasked with leading, deciding, and overseeing the AI governance strategy. More precisely, the AI governance committee works with the business and technical experts to define, review, monitor, and improve controls relevant to each phase of the AI development process. Moreover, the committee, through controls, aims to drive AI awareness and culture change within the organization wherever necessary.

Once defined, the development team ensures that appropriate controls, as defined by the AI governance committee, are implemented at every stage of the AI development process. As the organization continues to evolve and becomes more successful at using data strategically, conflicts in relation to existing controls may arise. These should be reported to the AI governance committee, which should review and decide on the situation and advise the development team accordingly. Sources of conflicts may include, among others, new requirements from strategic changes, expansion to other regions of the world, and the acquisition of another organization. As part of the overall governance strategy, the AI governance committee should ensure that the development team has the right skill sets to implement controls and is provided with appropriate training and mentoring wherever necessary. This is needed because they cannot be held accountable if not qualified to execute the task they are accountable for. Moreover, investment in new tools (including software and infrastructure, and processes) may be required to ensure that certain controls meet the expected standards.

Assessment of AI Governance Arrangements

AI governance strategy does not stop with delegating oversight of core tasks. Importantly, an organization also needs to be able to assess the effectiveness of its AI governance arrangements. How else does an organization know its governance strategy is working? As was discussed in the "AI Development Process Controls" section, the SAIF framework, through its controls element, provides an organization with the ability to clearly define characteristics that its AI system must meet. Most of these characteristics can be linked to practical metrics such as privacy, fairness, security, transparency, etc. By introducing these or similar metrics among its standardized organization-wide evaluation criteria and assessing these regularly, an organization can efficiently measure the effectiveness of its AI governance arrangements.

Considering fairness, for example, we discussed model monitoring as part of controls for the model deployment phase. An organization can periodically assess how well it is performing against the fairness metric by examining the model's output over the identified period of time and adjusting its AI governance arrangement and practice accordingly. For metrics such as privacy and security, one approach to access the effectiveness of the organization's privacy and security arrangement consists of monitoring and reporting privacy and security issues over identified periods of time. An increase or lack of improvement with respect to those metrics indicates a need to revisit the governance strategy in place.

In addition to an organization's internal mechanism for assessing its AI governance arrangements, and in order to introduce a further element of external accountability, an organization may rely on a regular audit by a third-party organization to assess the effectiveness of its AI governance arrangements.

Conclusion

In this chapter, we presented our proposed process – controls – governance model for the design, development, and deployment of AI systems. This forms the core element of the SAIF operating model and allows an organization to improve the understanding of its AI-related activities together with the associated risk profile and exposure to unwanted consequences. The "process" methodology is designed around the cross-industry standard process for data mining, which we enhanced by introducing a performance metrics phase that allows the integration of business requirements in the process. We further extended the development process with controls in the form of constraints on the "process" to ensure each stage is designed to yield the intended outcomes. Finally, we introduced governance through which an organization empowers its stakeholders for the implementation of the development process and controls.

AI Performance Measurement: Think Business Values

The film *Minority Report* starring Tom Cruise is one of the few movies that explore the potential impact of technology on everyday life. Beside its idealistic view of the future ahead of us, the sci-fi movie features an almost crime-free future where a special police unit known as the "pre-crime department" identifies and arrests criminals based on foreknowledge provided by "precogs."[1]

---

[1] The term precog is used to refer to people capable of precognition.

© Ghislain Landry Tsafack Chetsa 2021
G. L. Tsafack Chetsa, *Towards Sustainable Artificial Intelligence*,
https://doi.org/10.1007/978-1-4842-7214-5_5

For the sake of facilitating the understanding of the point we illustrate through this movie, the following is a short overview of the movie's main story:

> The movie is set in 2054, when a Washington, DC's prototype pre-crime police department stops murderers before they act, reducing the murder rate to zero percent. Murders are predicted using precogs, who "previsualize" crimes by receiving visions of the future. The federal government is in the process of adopting the program nationwide when John Anderton, the head of the pre-crime unit, learns he will murder a man he does not know in 36 hours. John immediately flees as agent Witwer, assigned by the department of justice to audit the program, begins a manhunt. Anderton seeks the advice of Dr. Iris, the creator of the pre-crime technology. Dr. Iris reveals that one of the precogs does not always agree with the other two, a "minority report" of a possible alternate future. Although known by Dr. Iris and Burgess, the director and founder of pre-crime, this information has been concealed from the public and the pre-crime police department programs.

Whether arresting criminals for crimes they will never get to commit makes sense or not is a discussion for another day. Let us instead concentrate on the use of technology from a performance perspective. In this respect, the film raises fundamental issues related to the development and widespread use of digital technology. More precisely, it highlights the problem of

- Setting unrealistic expectations for a system, which in this case made it impossible for the development team to develop a system that would meet expectations

- The lack of transparency and understanding of today's cutting-edge digital technologies as far as decision makers are concerned

For example, users of the pre-crime technology have no other way of evaluating/assessing the technology beyond its predictions and the steadfastly belief that it is right and fair. They are relying on a system of whose performance they have little to no understanding in order to make life-changing decisions about supposed criminals. While fictional, the pre-crime technology central to the film is in many ways not too dissimilar from many of today's mainstream technologies. For example, a loan approval prediction algorithm is often used as the sole base for accepting or rejecting loan applications. However, it is impossible to be certain about making the right decision about an individual's willingness or ability to repay given that individuals denied loans never get the opportunity to show that they could have been lending worthy.

Another example of an AI system rolled out to the public without adequate information on its evaluation and effectiveness is the Babylon Health AI system adopted by the UK National Health Service. For some context, Babylon's AI system analyzes a patient's input along with their answers to a set of questions to provide them with relevant health and triage information. On Babylon AI system's performance, a publication from *The Lancet* states

> *[…] It is not possible to determine how well the Babylon Diagnostic and Triage System would perform on a broader randomised set of cases or with data entered by patients instead of doctors. Babylon's study does not offer convincing evidence that its Babylon Diagnostic and Triage System can perform better than doctors in any realistic situation, and there is a possibility that it might perform significantly worse. If this study is the only evidence for the performance of the Babylon Diagnostic and Triage System, then it appears to be early in stage 2 of the STEAD framework (preclinical). Further clinical evaluation is necessary to ensure confidence in patient safety. […]* (Hamish, Enrico, and David 2018)

Although it is unrealistic to claim a prediction from a system to be certain, quantifying and documenting its performance on the basis of past performance helps to reduce this knowledge gap and helps to align the expectations between users of the system and the business and the development teams. DS practitioners generally have a good understanding of how to evaluate systems they develop, but often fail to clearly communicate the implications and meaning of reported performance metrics to business leaders. This is especially true when a reported performance metric does not tell how well the system really performs. For example, an AI system that predicts a deadly but curable medical condition that afflicts 0.01% of the population with an accuracy of 99.99% may still be a poor system. In this case, the system just needs to predict "not sick" irrespective of its input to achieve such accuracy. While an accuracy of 99.99% sounds exceptionally good, it does not account for the risk of leaving patients to die when they are mistakenly predicted "not sick." It is, therefore, not surprising that performance metrics used by the business and the development team are often misaligned. This points to the importance of having the development team and business working together closely to guarantee that the system under development or in use is optimized for the right objectives.

In this chapter, we discuss AI performance metrics and examine the need to involve business decision makers in the process of defining performance measurement metrics of AI systems. The remainder of the chapter is structured as follows:

- We first provide an overview of traditional approaches for evaluating AI systems.

- We then extend traditional metrics, for evaluating AI models, to include metrics that incorporate the organization's way of doing business and values, which we earlier referred to as soft performance metrics.

- Finally, we make some concluding remarks.

AI Performance Metrics Overview

All digital technologies have some limitations on how well they perform tasks they are designed for, with AI technologies being no exception. When looking at the performance of an AI system, there are multiple areas of interest, ranging from the performance of the AI algorithm, across system performance, to soft performance. This section provides an overview of performance metrics for AI algorithms (AI algorithm and ML algorithm are used interchangeably). The discussion of system performance, which refers to performance of the overall AI system,[2] lies outside of the scope of this book. Instead, the concept of soft performance, which is discussed in detail in the next section, involves performance metrics independent of the AI algorithm's ability to deliver expected outcomes. For example, assuming privacy is a concern, an AI system that violates the privacy of data subjects typically performs poorly on the soft performance scale, unrelated to the system performance.

ML algorithms can roughly be grouped into supervised and unsupervised algorithms depending on whether the input data is labeled or not. Supervised algorithms use labeled data to learn a systematic approach to approximating the label given to the input, whereas unsupervised algorithms attempt to find patterns and make sense of unlabeled data. In other words, supervised algorithms are generally used for "supervised problems," meaning problems where we have a complete sample with both information on the predictors and the variable to be predicted. In contrast, unsupervised algorithms are used for "unsupervised problems," for which we do not have example answers. A natural extension of supervised and unsupervised problems is the class of semi-supervised problems where the algorithm is provided with both labeled and unlabeled data.

These groups of problems can be illustrated using practical examples. Given attributes of ten households, for example, the number of bedrooms and occupants, and *information about the electricity consumption of each household*, the task of estimating the electricity consumption of a new household, for whom we also have full information on attributes, is a supervised problem. In this task, an example would be attributes of a household together with its actual electricity consumption. Now, assume that those households belong to three age groups and we do not know which households belong to a given age group. The task of dividing these households into three groups, so that households in each group belong to the same age group, is an example of an unsupervised problem.

The following reviews performance metrics for each of these classes of problems.

---

[2]Includes performance metrics applicable to any digital system.

Supervised Problems

The choice of metric(s) for assessing the performance of a machine learning model generally depends on the application at hand. For classification problems, which require the prediction of a class, category, or label, the most popular metrics include *classification accuracy* (commonly referred to as *accuracy*), *logarithmic loss*, *confusion matrix*, *area under curve (AUC)*, and *F-measure*. For regression problems, which require the prediction of a quantity, *mean absolute error* and *root mean square error* are most commonly used. Note that these metrics are meaningful only when they are computed on a test data set, that is, a data set that was not used for the initial computation of predictors or for training the model.

Before moving on, it is important to elaborate on some of these concepts through an example, such as credit card fraud. Credit card fraud is a problem commonly faced by high street banks. Consequently, they devote a considerable amount of effort to combatting credit card fraud. In order to support these efforts, banks could work with their DS department, who can use AI to actively monitor and identify suspicious or fraudulent credit card transactions as they take place. According to our earlier definition, this problem can be treated as a classification problem because it boils down to deciding whether a transaction falls under one of the following categories: fraudulent and non-fraudulent. Alternatively, for the sake of clarity, if we were concerned with the amount of money taken from an account at every transaction, then the problem would be treated as a regression problem. During the development phase, the DS team needs to assess how good their machine learning algorithm is at determining whether a transaction is fraudulent or non-fraudulent. To achieve this kind of assessment, it is regular practice to rely on evaluation metrics, which we discuss below. It is not necessary to understand them in order to continue following this book, so it is possible to jump to the "Summary of Performance Metrics for Supervised Problems" section for a short summary.

Classification Problem's Evaluation Metrics

This section highlights performance metrics most commonly used for the evaluation of classification problems:

- **Classification accuracy:** Generally referred to as accuracy, it is the proportion of correct predictions. This metric is generally not a very good indicator of the model's performance. For example, in the case of a fraud detection service with very few 0.2% fraudulent transactions, one can usually achieve a very high accuracy by simply classifying all transactions as non-fraudulent, given the small proportion of fraudulent transactions. This defeats the purpose of identifying fraudulent transactions.

- **Logarithmic loss:** Also known as log loss, it quantifies the accuracy of a classifier (classification algorithm) by penalizing incorrect classifications. It is particularly useful in settings where there are more than two classes, also referred to as multiclass classification. The lowest value the log loss can take is zero (0), in which case the classifier is perfect. As a rule, a value close to zero indicates high accuracy, while a value far from zero indicates poor accuracy.

- **Confusion matrix:** Often referred to as error matrix, a confusion matrix on its own is not an evaluation metric, but a convenient tabular structure to capture and visualize the performance of a classification model. In addition, it is the basis for other metrics discussed below. Using the fraud detection problem introduced earlier, let us assume we have a classifier which takes as input 28,925 transactions of which we know the true label. The classifier's job is to label each transaction from the sample as either *fraudulent* or *non-fraudulent*. We can compare the performance of our classifier to the true label using a confusion matrix as illustrated by Table 5-1, where the values along the diagonal running from the top-left cell to the bottom right indicate the number of transactions that were correctly classified.

Table 5-1. Confusion Matrix for the Fraud Detection Classifier

		Predicted	
		Non-fraudulent	**Fraudulent**
Actual	**Non-fraudulent**	(TN)	(FP)
		28,158	274
	Fraudulent	(FN)	(TP)
		92	400

The following concepts are generally associated with a confusion matrix:

- **True positive (TP):** The observation is fraudulent (positive) and is predicted fraudulent.

- **True negative (TN):** The observation is negative (non-fraudulent) and is predicted negative.

- **False positive (FP):** The observation is negative (non-fraudulent) but is predicted positive (fraudulent).

- **False negative (FN):** The observation is positive (fraudulent) but is predicted negative (non-fraudulent).

Here, fraudulent is positive and non-fraudulent is negative.

False positive and false negative as highlighted in Table 5-1 are of particular interest because they inform on the algorithm's potential to make the wrong decision. Depending on the nature of the problem at hand, false positives may outweigh false negatives or vice versa. Using the fraud detection example, it is probably better to have false positive than false negative.

Area Under Curve (AUC)

Another metric for assessing the quality of a classifier is the area under curve. Ranging from zero to one, the area under the curve is used to assess the ability of a binary classifier to discriminate between positive and negative observations. The higher the AUC, the better the model is at classifying negative observations as negative and positive observations as positive. The curve referred to here is the receiver operating characteristic (ROC) curve where the true positive rate (sensitivity or recall) is repeatedly plotted against the corresponding false positive rate (1-specificity), derived by estimating the classifier repeatedly while varying the classifier parameters. The terms involved in the definition of the AUC are defined as follows:

- **True positive rate/recall/sensitivity:** Also known as sensitivity or recall, it corresponds to the proportion of positive observations that are correctly classified as positive. It can be thought of as the probability that an observation will be classified as positive given it is actually positive. Using the fraud detection example, it is the probability that a transaction will be classified as fraudulent when it is fraudulent. Formally, it is defined as illustrated by Equation 5.1. Recall emphasizes on false negative, and a high recall usually indicates low false negative.

$$Sensitivity\ /\ Recall\ /\ True\ positive\ rate = \frac{TP}{\left(TP + FN\right)} \tag{5.1}$$

- **Specificity:** Specificity corresponds to the probability that an observation will be classified as negative when it is actually negative. Following our fraud detection example, it is the probability that a transaction will be labeled as non-fraudulent given it is non-fraudulent. The formal definition of specificity is provided by Equation 5.2.

$$Specificity = \frac{TN}{(FP + TN)} \qquad (5.2)$$

- **False positive rate (FPR):** It corresponds to the proportion of negative examples that are incorrectly classified as positive and can be derived from specificity as illustrated by Equation 5.3

$$False\ positive\ rate = 1 - Specificity \qquad (5.3)$$

Figure 5-1 illustrates the AUC for the fraud detection problem. The diagonal straight line depicts the performance of a random classifier. That is a classifier that randomly classifies a transaction as fraudulent or non-fraudulent and has no predictive ability. A classifier that falls below the diagonal is a bad classifier because it performs no better than random, whereas a classifier that falls above the diagonal is generally acceptable depending on the problem. The optimal case is where the AUC tracks the vertical axis upward, and the true positive rate remains to 1 across various false positive rates.

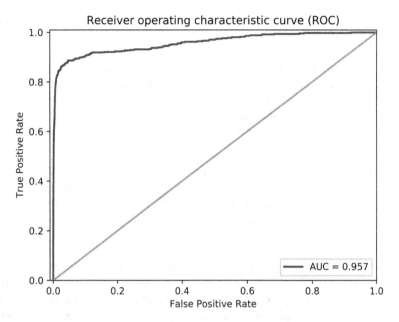

Figure 5-1. AUC-ROC curve for the fraud detection classification task

Choosing the "Right" Performance Metric for Classification

One may wonder why there are so many metrics to choose from. The choice of the "right" metric often depends on the problem one is trying to solve and the level of understanding of the application domain. This underscores the necessity for the DS team to work hand in hand with the business. To illustrate this, consider the example of a classifier used for screening patients for cancer. Let us assume that classifier has the following performance characteristics:

- Sensitivity = 0.90
- Specificity = 0.85
- False positive fraction = 0.15
- False negative fraction = 0.10

Therefore

- If a patient has cancer, there is a 90% probability that the screening test will be positive (classifier predicts the patient has cancer).
- If the patient does not have cancer, there is an 85% probability that the screening test will be negative (the classifier predicts the patient is not diseased).

The error of the screening test is quantified by the false positive and false negative fractions.

- If a patient does not have cancer, there is a 15% probability that the test will be positive. Roughly speaking, if the patient does not have cancer, there is 15% probability that the classifier will incorrectly label the patient as having cancer. This can be of great concern, as a positive result would likely create anxiety.

- If the patient is suffering from cancer, there is a 10% probability that the test will be negative. In other words, if a patient has cancer, there is a 10% probability that the classifier will label the patient as not having cancer, giving the patient and their family a false sense of assurance.

This suggests that the most appropriate performance metric, as illustrated above, may depend on the implications of an error, highlighting the need to consider performance metrics in the context of business objectives. In the case of the cancer example above, a false negative is more troublesome, while a false positive at least allows the patient to take more tests to understand if it was a true or false positive test result. So, we would likely want to reduce the likelihood of false negatives, even if this means an increase in false positives.

Note In clinical settings, a confidence interval (CI) is often provided alongside specificity and sensitivity. Formally, the CI proposes a range of plausible values for an unknown estimator. Associated to this CI is the confidence level that the true parameter is in the proposed range.

While evaluating the performance of an algorithm, tracking multiple metrics like sensitivity and specificity can be difficult as one often has to make a trade-off. So, it is often convenient to use a metric that combines multiple individual metrics into one metric. One such metric is the F-measure discussed below.

F-Measure

Also known as F-score or F1-score, the F-measure is typically used for binary classification and relies on two key concepts: *recall*, which we introduced earlier, and *precision* introduced below.

Precision or positive predictive value (PPV) is the proportion of observations that are correctly classified as positive from all observations that are classified as positive. In other words, the probability of a subject with positive predictive value to truly belong to the positive class. Formally, it is defined as in Equation 5.4:

$$Precision = \frac{TP}{\left(FP + TP\right)} \qquad (5.4)$$

While precision penalizes classifying an observation as positive when it is not (false positive), recall penalizes not classifying an observation as positive when it is (false negative). In other words, high precision is preferable when the cost of false positive is high, whereas high recall is preferable when the cost of false negative is high. Using our fraud detection example, a high recall is preferable because it is better to wrongly identify a transaction as fraudulent, allowing further checks to be undertaken, rather than not identifying it and thus incorrectly missing a potentially fraudulent transaction.

It can be difficult to track both precision and recall at the same time, especially during the development phase. This is where the F-score comes into play. As illustrated by Equation 5.5, it is defined as the harmonic average[3] of precision and recall. It can be thought of as some sort of balance between precision and recall. One interesting property of the F-measure, which is inherent to the harmonic mean, is that it lies closer to whichever metric between precision and recall is smaller. For example, if precision is smaller than recall, the inverse of precision will be bigger than the inverse of recall. Consequently, the resulting F-score will be closer to precision than it is to recall.

$$F\text{-}measure = \frac{2 * precision}{\left(precision + recall\right)} \qquad (5.5)$$

Regression Problems' Evaluation Metrics

As highlighted earlier, the most popular performance metrics for regression problems involve the root mean square error and mean absolute error. The lower these metrics, the better is the model.

- **Mean absolute error (MAE):** The MAE measures the absolute magnitude of the difference between predicted and the actual values. It is the average of the absolute differences between prediction and actual observation. The MAE is formally defined by Equation 5.6:

$$MAE = \frac{1}{N} \sum_{1}^{N} \left| y_i - \hat{y}_i \right| \qquad (5.6)$$

---

[3]Reciprocal of the arithmetic mean of the reciprocals of the given set of observations (Wikipedia). For example, given three numbers 2, 3, and 5, the average of their reciprocal is given by (1/2 + 1/3 + 1/5) /3 = 1.03/3, leading to a harmonic mean of 3/1.03 = 2.91.

where N is the number of observations, y_i the value of observation i, and $\hat{y}_i$ the estimated/predicted value for observation i.

- **Root mean square error (RMSE):** Similar to the MAE, the RMSE is used to measure the absolute difference between predicted values and actual (or measured) values. It is computed as illustrated by Equation 5.7. Note that the differences are squared before averaging. As a result, a large error contributes more strongly to the RMSE than a smaller error. Consequently, the RMSE might be more appropriate when large errors are undesirable. However, it is less easily interpretable than the MAE.

For example, given an AI model that estimates the yearly energy consumption of households in London with a root mean square error of 4, can we actually say anything about the algorithm's performance? This doesn't say much about actual estimates. Consequently, in this case the root mean square error is a good metric for developing the model, but not necessarily a good metric for reporting the performance of the model to the business.

$$RMSE = \sqrt{\frac{1}{N}\sum_{1}^{N}\left(y_i - \hat{y}_i\right)^2} \tag{5.7}$$

Here, N is the number of observations, y_i the value of observation i, and $\hat{y}_i$ the estimated/predicted value for observation i.

Summary of Performance Metrics for Supervised Problems

The common feature among performance metrics for supervised learning problems is that they all have strong mathematical foundations. This is very useful as it provides developers with a framework to develop and evaluate models with minimal interference from business decision makers. The flip side to this is that business decision makers often have little to no understanding of what these metrics translate to as far as their day-to-day business operations are concerned. This impacts the usefulness of performance metrics in helping the DS team develop a model that best meets business objectives.

Unsupervised Problems

Common unsupervised problems can generally be grouped into clustering and dimensionality reduction and association. Clustering aims to discover groups inherent to the data. A practical application of this technique is customer segmentation, where a service provider attempts to segment its customers into groups such that customers in each segment have some things in common (e.g., electricity consumption pattern, response to an ad campaign). Association analysis attempts to discover interesting relationship rules that describe a large proportion of the data. For example, a utility company would be interested in knowing what other of its products customers are likely to buy when they order a boiler installation. This information can then be used to bundle multiple products together or suggests the other products when they order a boiler installation. The most common application of dimensionality reduction is data visualization. This technique attempts to find a new representation of the data such that as much information about the original data is preserved. It provides substantial benefit because one can reduce the dimension of the data to two or three coordinates for better visualization. Dimension reduction is widely used by biologists as a preprocessing step (e.g., visualizing high-dimensional genomic data) for subsequent analysis.

Assessing the performance of unsupervised problems is a difficult task. This is attributed to the fact that for unsupervised learning problems, there is generally no reference "ground truth" outcome to compare once proposed solution against. For clustering, key approaches include internal, external, manual, and indirect evaluation (Feldman and Sanger 2006):

- Internal evaluation summarizes the clustering into a score reflecting its quality.

- With external evaluation, the clustering is compared to a "ground truth" classification.

- In manual evaluation settings, the clustering is inspected by a human expert.

- Indirect evaluation consists of evaluating the utility of the clustering in its intended application. In this setting, clustering is considered a step to help perform another task. The appropriate clustering is then assessed by looking at the performance of the other task. For example, in a demand response program,[4] the primary task is to get users to shift their electricity usage during peak periods to off-peak periods. Users can generally be clustered into different segments based on their electricity usage patterns. Allowing for the identification of user segments suitable for the demand response program. A few clustering methods can be tried, and their results are evaluated based on how well they help with the demand response task.

---

[4]Gives customers the opportunity to improve the efficiency of the electric grid by shifting their usage during peak hours to off-peak hours.

Unlike manual and indirect evaluation methods, internal and external evaluation approaches are rarely used to judge a clustering in practical applications. Nevertheless, they provide useful statistics for identifying bad clustering (Weiss et al. 2004).

In association analysis settings, the emphasis is on selecting interesting rules from the set of possible rules (Hipp, Guntzer, and Nakhaeizadeh 2000; Hornik, Grün, and Hahsler 2005). This is accomplished via constraints on various measures of significance and interest. Among these, the most commonly used are minimum thresholds on support and confidence. The "support" of an itemset[5] can be thought of as the frequency of an itemset in a data set. Given two itemsets and an association rule linking them, the "confidence" of the rule is an indication of how often the rule has been found to be true. For example, consider the case of a supermarket with the following itemsets {*milk, bread*} and {*butter*} and the association rule {*milk, bread*} => {*butter*} (meaning in simple terms that a customer who buys bread and milk will likely buy butter as well). The support of the itemset {*milk, bread*} is the proportion of transactions containing milk and bread. Let us then imagine that we calculate the confidence of our example association rule to be 0.5. This means that for 50% of transactions containing milk and bread, the rule is correct (Hornik, Grün, and Hahsler 2005).

Practical applications of unsupervised problems tend to be aligned with business expectations or at least well understood by business decision makers. This is because the ultimate goal of unsupervised learning is to create either clusters (in the case of clustering) or rules (in the case of association learning) that are meaningful in the intended application domain. Roughly speaking, real-life applications of unsupervised problems are generally embraced by the business or domain experts because

- They are often consulted during the validation phase, because the widely used indirect and manual evaluation methods require input from domain and business experts.

- And the story told by the outcome is often something familiar that people can relate to or has a clear business implication.

---

[5]Group of related elements.

Beyond Traditional AI Performance Metrics

In this section, we discuss soft performance metrics and highlight the need to formulate traditional AI performance metrics as business objectives through a use case.

Soft Performance Metrics

As discussed earlier in this chapter, AI performance metrics can be both complex and often disjointed from business objectives. Chapter 4 advocated for traditional AI performance metrics to be extended to include a new class of metrics which we refer to as soft performance metrics.

Soft performance metrics can be thought of as constraints that are imposed upon the AI system throughout its life cycle. Set constraints may vary greatly from one organization to another. Consequently, we limit our discussion to those involved in building blocks for sustainable development and use of AI. These include, but are not limited to, bias, privacy, and interpretability. Given the multitude of approaches to defining these concepts, organizations need to maintain an organization-wide definition and interpretation of each of these concepts wherever standards are lacking.

Bias

Also referred to as algorithmic bias, it occurs when the outcome of an algorithm is unfair,[6] generally to a specific group of individuals. There is no standard approach for evaluating an AI system against bias. Nevertheless, one can test for bias by scrutinizing and monitoring the data used by the algorithm and its outcome for unwanted behavior throughout the DS development process. Data is arguably the biggest contributor to bias in AI, which can be attributed to the fact that AI algorithms fundamentally depend on input data (knowledge base) during the development phase (training) to build their decision-making logic. Consequently, the algorithm's decision-making, irrespective of the problem at hand, inherits flaws and imperfections inherent to the data.

Unveiling flaws and imperfections inherent to the data involves understanding the target audience[7] of the AI system. This helps to identify attributes of the audience that may represent a source of bias; we will refer to such attributes as "sensitive attributes." Sensitive attributes usually include most demographic

---

[6]Such as privileging an arbitrary group of users over others. For example, current state-of-the-art facial recognition systems have a higher error rate when attempting to identify dark-skinned and female faces.

[7]The target audience is the population of interest.

information such as sex, occupation, age, etc. One can then monitor the output of the system (predictions and recommendations) for bias given those attributes. Approaches for mitigating bias include, but are not limited to

- Involving end users of the system and business decision makers earlier in the development process.

- Adjusting the classifier such that all groups have (almost) equal probability of being assigned the positive outcome for classification problems. Similarly, one should seek a balanced prediction error across all groups in a regression setting.

- Ensuring that all groups have similar false positive and false negative rates.

- Making sure that the model is adequate for the underlying problem. Using the problem of estimating a household electricity consumption, one could decide to create a model that simply estimates a household's electricity consumption as the average consumption of all households on the same street. Such a model is inadequate for the problem because each household possesses individual characteristics that are generally not captured by averaging the consumption of households on a street. As a result, the prediction error will be significantly larger than using a model that takes into account households' individual characteristics.

- Balancing sensitive attributes. For example, a facial recognition system that is trained primarily on male would have difficulties identifying females. Using an almost equal number of males and females during the development process can help the algorithm learn to identify both groups equally.

As discussed earlier, not having enough data points from certain groups of our target population can lead to bias. Data augmentation techniques such as oversampling and methodologies such as stratified sampling could help reduce bias (Vasileios and Eirini 2018). In simple terms, oversampling (more precisely, random oversampling) consists of creating a data set that has as many instances of the minority group as instances of the majority group. This is achieved by duplicating random instances of the minority group. Synthetic Minority Oversampling Technique (SMOTE) (Chawla et al. 2002) is another popular technique designed to improve random oversampling for classification problems. Unlike random oversampling, SMOTE does not duplicate instances of the minority group, instead, as its name indicates, it creates synthetic instances of the minority group. More precisely, SMOTE oversampling works by picking each instance of the minority group and finding its k-nearest neighbors. It then creates a synthetic instance along the lines joining the minority instance and any or all of its neighbors. Despite its success, some research suggests that SMOTE may not have a significant impact when applied to large data sets (Blagus and Lusa 2013).

Techniques such as stratified random sampling can be used to ensure that each subgroup within the target population receives appropriate representation. However, it is not always possible to use stratified sampling, especially in circumstances where some instances fall into multiple subgroups. This is because stratified random sampling divides the population of interest into smaller groups (known as strata), based on shared characteristics.

Stratified k-fold cross-validation is another approach for addressing issues surrounding bias. Typically used when the size of the data set is small, k-fold cross-validation is a model validation technique. In the k-fold cross-validation setting, the data is randomly split into k partitions or folds of equal size. One of the k samples is retained as the validation data set, while the remaining k-1 are used to train the model. This process is repeated k times with each partition used exactly one time as the validation data set. As a result, a total of k models are created and validated. These models' performance are then averaged to produce a single estimate of the model's performance. Although k-fold cross-validation generally produces a good estimate of the model's performance, it is not as effective when the data set has an imbalanced class distribution. This is because certain partitions created during the k-fold cross-validation process may not reflect the distribution of the original data set. In other words, some partitions may have too few or no examples of the minority class. Stratified k-fold cross-validation solves this problem by ensuring that the class distribution in each partition of the k-fold process matches the distribution of the original data set (He and Ma 2013). In this setting, the model needs to predict both the minority and the majority classes correctly to achieve good performance, as opposed to only having to correctly predict observations of the majority class.

More recently, the AI community has been actively investigating approaches to efficiently identify and reduce or even remove algorithmic bias. The AI Fairness 360 open source toolkit is an illustration of such efforts.[8] Nevertheless, involving a wider and diverse audience in the development of AI systems probably remains a sensible approach to identifying and mitigating bias. The rationale behind this is that humans usually tend to add more context than algorithms to their decision-making. For example, an AI system that predicts whether a patient has tuberculosis from a chest X-ray uses a fixed context which in this case is the chest X-ray. However, medical doctors tasked to do

---

[8]https://aif360.mybluemix.net

the same would likely incorporate more information such as balancing symptoms with side effects of other actions the patient may have taken to reduce the symptoms.

Extra information for an AI system can be captured through graph representation learning (Bo et al. 2018; Thomas and Welling 2016) or graph embeddings. The latter can then be constrained to enforce fairness (Avishek and Hamilton 2019) and mitigate bias. Similarly, other in-processing (during training) approaches for mitigating bias involve using special techniques such as adversarial machine learning (Beutel et al. 2017; Hu Zhang, Blake, and Mitchell 2018) or applying constraints to the algorithm so that it exhibits some form of fairness (Kamishima, Akaho, and Sakuma 2011; Muhammad, Valera, and Manuel 2015).

When discussing bias, it is natural to wonder whether throwing sensitive attributes out of the data set may solve the problem. Unfortunately, this is not sufficient because many attributes often correlate with sensitive demographic information (Kamishima, Akaho, and Sakuma 2011; Feldman et al. 2014; Aditya and Robert 2017).

To summarize, it is important to understand that algorithmic bias may give rise to unfair outcomes, for example, in the form of discrimination. That said, it is probably impossible to have a bias-free AI system given current data sources. As a result, identifying and documenting limitations caused by algorithmic bias is already a great step forward. Since algorithmic bias is inherent to our society and, therefore, the current knowledge base (data), this may mean that fully removing algorithm bias ultimately requires discarding existing data we have so far collected. In other words, data currently sitting in most organizations' databases is of poor quality and inadequate for a (near) bias-free data-driven strategy. As a result, when aiming to develop a near bias-free data-driven strategy, organizations will need to define problems they want to address and design the associated data collection accordingly. Algorithms are deployed in technologies we use in healthcare, education, government, criminal justice, and economic systems. As a result, algorithmic bias can impact operations at a society's institutional level, effectively acting as a social force with a very wide reach. Reducing and ultimately removing algorithmic biases from applications should be a key priority of all future AI developments to avoid perpetuating existing biases.

Privacy

Privacy is a complex concept to grasp. The introduction of data protection regulations such as GDPR has highlighted a core privacy aspect, namely, the exercise of control over users' personal data. Yet, from an organization's perspective, meeting the need of control demanded by users is even more complex than establishing the concept itself. This is because very sensitive personal attributes can accurately be inferred from easily accessible digital records of behavior such as social media likes (Kosinski, Stillwell, and Graepel 2013). Consequently, organizations need to develop and implement privacy policies beyond data protection regulations in order to meet users' expectations. Such an exercise requires organizations to firstly develop an understanding of the factors affecting users demand for privacy and to secondly design and implement policies to mitigate privacy concerns. This is of particular importance when organizations use a third party for computation (e.g., using machine learning services through an application programming interface (API)) or are sharing user data.

Some organizations naturally anonymize user data by removing personal identifiers before sharing (either with other organizations or the public). However, as argued above, de-anonymization is often possible given the widespread availability of background knowledge information, meaning that anonymization by itself is usually not enough to preserve user privacy. For example, using their knowledge of the Internet Movie Database (IMDb) and anonymized data released by Netflix for the Netflix movie recommendation competition, Narayanan and Shmatikov (Narayanan and Shmatikov 2008) managed to identify Netflix records of known users and further inferred their apparent political preferences.

Alternative approaches for developing machine learning algorithms could be considered instead. For example, privacy-preserving algorithms can allow multiple parties to jointly build a machine learning model without having to share their input data set (Shokri and Shmatikov 2015; Bonawitz et al. 2019; Tang et al. 2019; Phong et al. 2018). The TensorFlow Federated framework, developed by Google, is an attempt to make privacy-preserving algorithms available to the wider AI community. Despite their potential, privacy-preserving technologies present specific challenges related to scalability and security assumptions (Al-Rubaie and Chang 2019). In addition, necessary precautions need to be taken to avoid bias and should ideally be coded into the privacy-preserving learning framework.

While most of the discussion about privacy is relevant to both AI and non-AI systems, the latter discussion demands more attention because of

- The need for AI systems to collect and aggregate more data

- The ability of AI systems to accurately infer information that was not explicitly disclosed through the data collection and aggregation process

To illustrate, let us assume an organization collects data about its customers (whether with or without their consent); there are two main scenarios: (a) the data is used by an AI system and (b) the data is not used by an AI system. In the latter scenario, the organization could be using the data for computing statistics about its customer base, which one can think of as "facts" about the organization's customer base. Doing so has limited impact on the privacy of any given individual because those facts are explicitly available from the data.

In the former scenario, the AI system has the potential, through its internal processes, to create "new"[9] information from the data. This is undesirable in some cases because of the nature of the new information and/or the implications that it has on the decision that the AI system makes. For example, the system may accurately predict the user's race, which the user may not have been happy to disclose, and may further make decisions on the basis of this information which the user may not agree with. The affected user, even if they provided consent for the data to be collected, rarely explicitly consents to the organization using or having such new information. Note however that an organization may still be liable for the misuse of the inferred information even when it is false or inaccurate.

Interpretability

The concept of interpretability has long been, and still remains, a major concern in AI. This can be attributed to the success of AI and its ability to impact almost every aspect of our life. There is no standard definition of the concept of interpretability in machine learning; however, interpretability is generally judged by how well people, such as users, understand the decision-making process of a machine learning model (Miller 2017). For example, simpler models such as linear regression and decision trees are naturally more interpretable than complex models such as neural networks. Conversely, complex models are generally more accurate than their simpler counterparts. As a result, accuracy and interpretability may often represent conflicting objectives, and one needs to find the right balance between the two.

Because machine learning models inherit bias from the data, making a model more interpretable by making it easier to explain the decision-making of a model can help one to assess the degree to which the model exhibits the following traits (Finale and Been 2017):

- **Fairness:** As discussed earlier, the model should not explicitly or implicitly discriminate against a given group of people and instead aim for unbiased predictions. For example, an interpretable model could transparently explain why it has decided that an individual is at high risk of committing a crime, allowing humans to assess whether the decision was based on sensitive attributes alone or an unbiased logic.

---

[9]New from the organization's perspective, because the customer who has provided the data may already hold that inferred information.

- **Privacy:** An interpretable model should make it easier to understand if sensitive information in the data is protected.

- **Reliability or robustness:** Small changes in the input should not lead to large changes in the output.

- **Causality:** The model relies on causal relationships rather than mere correlations, which would not imply predictive power.

- **Trust:** People tend to trust a system that they understand.

While certain machine learning models rely on data to derive new insights, some others are concerned with producing a new representation of the input data. Optical character recognition (OCR), an application of machine learning, performs the conversion of images of typed, handwritten, or printed text into machine-encoded text. As a result, it is an example of the latter category, whereas a model that assesses the risk of an individual committing crime falls within the former category. As a rule, models, such as OCR, that are concerned with producing a new representation of the data do not need to be interpretable because they simply aim to provide an "identical" representation of the information they are given. However, organizations should assess and document which of their AI systems need to be built around an interpretable machine learning model. In this regard, policy makers can help maintain consistency across organizations or industries by providing guidelines on applications that need to be interpretable.

Gillespie (Gillespie 2017) argues that applications where users can manipulate input to their advantage may not be interpretable. For example, in the fraud detection example introduced above, if a fraudulent user knew that random withdrawal of a large amount of cash increases the likelihood of them being detected, they may decide to make several withdrawals of relatively low cash amounts instead.

Typically, approaches for achieving interpretability consist of one of the following:

- Restraining the model's capacity to simple models. For example, using linear regression or decision trees.

- Relying on tools that analyze the model after the training phase is completed. These may, for example, reveal how much each attribute contributes to the algorithm's decision-making.

Involving the Business in AI Performance Metrics Design: From Performance Metrics to Business Objectives

This section emphasizes the need to involve the business[10] in the data science development process with a use case. Using the example of energy consumption estimation, this section illustrates how traditional AI performance metrics can be translated into business metrics that meet the organization's values and way of doing business. Beyond creating an AI system that is in line with the organization's values, a key benefit of translating traditional AI performance metrics into business or organizational values is that it creates a natural communication channel between parties involved in the development of the AI system. As illustrated below, this process may be influenced by various factors ranging from regulatory factors to human factors and to the organization seeking to maintain its reputation.

Energy Consumption Estimation

The following discusses the problem of estimating energy consumption for the direct debit customers of a UK energy provider. In order to set up a customer on Direct Debit (DD), energy providers generally calculate a 12-month estimate of their consumption, resulting in an annual quote for the customer's chosen tariff plan.[11] The amount of the monthly DD installment is then derived accordingly. However, there may be changes in the customer's circumstances causing their usage habits to go up or down. As a result, the amount derived in the abovementioned way may change. Consequently, a DD reassessment is generally performed after 6 months to ensure that the payment matches the customer's actual consumption.

It is now interesting to consider how the reassessment of the initial estimate and associated implications impacts the way of working of the data science team in charge of delivering the consumption calculator/estimator. Although any approach can be used for estimating users' electricity consumption, regulators often require utility companies operating in the UK to be able to explain the estimate to their end users. In addition, with the competition constantly increasing, sales operators need to be able to explain, for example, why a quote for a one-bedroom flat seems significantly higher than what they would expect. This calls for an accurate, yet interpretable, estimator, two objectives which, as highlighted earlier, may be conflicting.

Organizations naturally prioritize interpretable estimators over accurate energy estimators whose outcome cannot be explained. In general, accuracy

---

[10]Generally, any party with an interest in the outcome of the system.
[11]For simplicity, we suppose that the tariff is fixed for the estimation period.

may often be compromised in favor of similar business requirements. This of course depends on the nature of the task at hand; in general, a poor estimate that can be explained is preferred over a good one that cannot.

Additionally, the need for a DD reassessment imposes further constraints on how the data science evaluates the estimator. A DD reassessment typically reviews the initial assessment and either increases, reduces, or keeps the monthly installment amount the same. Because of our nature, most people will not question their provider if their monthly DD suddenly goes down but are more likely to do so if it goes up. From a business perspective, this incurs a handling cost for each customer call and potentially affects the provider's reputation or ability to retain customers. In other words, the business's interest lies in minimizing the number of calls it will potentially receive following the DD reassessment while also making sure they receive adequate and timely payment from the consumer. When this is not possible, the business would need a clear estimate of the proportion of customers likely to call following the DD reassessment.

As discussed earlier, the residual mean square error is a common approach for assessing regression problems like the one at hand. The challenge is then to create a metric that abstracts from the residual mean square error to provide an estimate of the number of potential callers following the DD reassessment 6 months later. We do not devise a complete solution in this book, as organizations can do so by themselves to gain strategic advantage over their competitors. In summary, the estimator, in addition to being accurate, needs to be interpretable and at the same time minimize the number of calls (which may bias estimates to overestimate the amount of electricity consumed). More generally, this section highlights the importance of deriving performance metrics from the business need.

Conclusion

This chapter discussed approaches for assessing the performance of AI systems. These approaches can be grouped into traditional and nontraditional approaches. Both have their own advantages and disadvantages. On the one hand, traditional approaches, typically used for supervised problems, are well established and understood by AI system designers. However, they are often disjointed to the organization's objectives. On the other hand, nontraditional metrics, typically used for unsupervised problems, are generally in line with the organization's expectations. However, they often lack a mathematical framework, are difficult to define, and are generally problem specific. Additionally, this chapter extends AI systems' performance metrics to include soft performance metrics and emphasizes the need to derive such AI performance evaluation metrics from objectives that reflect the organization's way of doing business wherever possible.

SAIF in Action: A Case Study

Financial institutions in the United States and Europe[1] are traditionally known to be biased against minority groups. Such bias is often observed in an organization's choice of who is offered a credit/loan or who is denied. Similarly, these minorities are often charged higher interest than their counterparts (Stefan et al. 2018; Bartlett et al. 2019; Aldén and Hammarstedt 2016; Lloyd, Bo, and John 2005; Solomon, Alper, and Philip 2013; Patatouka and Fasianos 2015). At the time of this writing, most of the leading financial institutions are doing little to nothing to combat the problem of systemic bias observed in the banking sector and more generally in access to finance. Additionally, financial institutions are increasingly relying on AI for improving operational efficiency through automation. Such reliance on AI technology is likely to amplify bias against minorities. This is because most if not all AI systems created in the banking sector as of today must rely on historical data that inevitably reflects in some way or form the current situation in the industry. Stated differently, the data is likely to incorporate the same flaws that one may be trying to

---

[1]This probably applies to other regions of the world as well; however, published studies examining the topic of financial discrimination against minorities are mostly focused on the United States and Europe.

G. L. Tsafack Chetsa, *Towards Sustainable Artificial Intelligence*,
https://doi.org/10.1007/978-1-4842-7214-5_6

combat. But more importantly, as discussed in the "Beyond Traditional AI Performance Metrics" section of Chapter 5, bias resulting from an algorithm's behavior operates at the institution's level and has a wider reach than bias exhibited by humans.

Considering what precedes, it is essential to ensure that the automation of the loan/credit decision-making process remains as fair as possible for applicants while maximizing the organization's profit. Maintaining this balance between business efficiency and fairness is of great importance because of the negative impact a misplaced decision may have on the applicant and in some cases the organization's reputation. Ironically, this is one of the very few instances in which it is virtually impossible to guarantee that applicants who are denied a loan/credit indeed deserved to be rejected. In other words, those who are denied a credit never get the opportunity to prove that they were credit worthy. This is unfortunate because getting rejected for a loan or credit may adversely affect one's creditability and future borrowing potential.

This chapter presents a practical implementation of the SAIF framework. This is achieved through the use case of a credit scoring system that relies on AI for its decision-making. Via this case study, we hope to provide the readers with an illustration of how the SAIF framework could be applied in their organization; however, it is important to point out that each organization needs to adapt the framework to its environment. The choice of the credit scoring case study is motivated by the reasons highlighted previously. More importantly, communities are increasingly in need of support from financial institutions, and not receiving this support because of misplaced decision-making would be unfortunate and may constitute a barrier to the development of those communities.

For the sake of clarity, throughout the remainder of this chapter we assume that the AI system is being developed for and by an organization which will be referred to as Financials Corp.

The remainder of this chapter is organized as follows:

- First, we provide some background information on the credit scoring use case.

- Next, we examine SAIF practices applied to the credit scoring system in the context of Financials Corp.

- Finally, we make some concluding remarks.

Background

To improve its operations and increase competitiveness in Europe, Financials Corp. has decided to streamline its mortgage and personal loan products by relying upon AI to help decide who gets approved for a loan. Following this decision, the organization (Financials Corp.) appointed a project manager and hired a team of data professionals to design and build the much-needed AI system. While this description may sound overly simplistic, it is in fact an accurate depiction of how a vast majority of AI-related projects are typically introduced or initiated in many organizations. In some cases, and more than often, it happens that the organization recruiting the team has no clarity of what the team is going to do. Meaning that the team probably needs to figure out what to do by itself.

To ensure that the day-to-day development of the new systems meets the organization's business objectives while preserving if not improving its reputation, Financials Corp. decided to adopt the SAIF framework.

A successful implementation of the SAIF in an organization takes effort and requires alignment from the entire organization. People are essential for its success, and involvement must start from the top of the organization with the chief executive officer (CEO). In the case of Financials Corp., key roles or people include the organization's CEO, the head of AI or chief data officer (CDO), the project manager who also acts as domain expert from the credit risk team, and finally the development team, which is composed of three data scientists/engineers and a team lead who is also the head of AI.

In this setting, and for the simplicity of this use case, we assume that the organization's CEO has defined the organization's value and vision, which can be summarized through the following statement:

> [...] Financial Corp. will not treat people differently on the basis of race, colour, age, disability, sexual orientation or identity [...].

The head of AI is empowered by the CEO not only to achieve this vision but also to maintain it through the system under development. This means that the organization is committed to supporting the implementation of SAIF not only by providing necessary resources but also by committing to understanding the benefits of the system under development and what is needed for its success. Such support from the organization's leadership is extremely important because it creates an environment through which the technical team can educate the leadership on the business benefits of the system under development and what it means for the organization.

Often, organizations are extremely slow, especially big organizations, meaning that requests from the development team to other departments may take months to be addressed. Worse, sometimes such requests may be related to the access to the data the team needs to get the job done. Having the support of the leadership team is usually a significant boost for the project, in the sense that other departments often find it easier to justify the allocation of their resources to addressing issues that the development team may face on a daily basis.

Beyond their role in leading the organization's AI strategy, the head of AI leads the development and is responsible for the project's success through the deployment of SAIF practices. In practical terms, this consists of defining appropriate controls for relevant stages of the DS process and overseeing their implementation. The head of AI achieves this by

- Collaborating with other departments such as the IT department, the legal, and the human resources departments

- Delegating responsibilities where appropriate

- Providing adequate support and training to the development team

The next section discusses some of the controls applicable to the design and development of the credit scoring system along with government arrangements required for the implementation of identified controls. These controls are defined on top of the data science development process developed in the "Data Science Development Process" section of Chapter 4.

Process – Controls – Governance for the Credit Risk Assessment System

Problem Formulation

The problem formulation phase aims to provide the development team with a concise and clear problem statement. In the case of the credit risk scoring, it is formulated as follows:

Using information (such as gender, age, …) submitted by applicants during their application creates a system to assess their credit worthiness. In other words, develop a credit scoring system that can be used to determine if an applicant is a bad credit risk or a good credit risk.

In technical terms, this is a classification problem consisting of classifying applicants into two user categories or classes, namely into bad credit and good credit applicants given their application details.

There are no controls for the problem formulation phase of the DS process; however, governance requires that the project manager (who also acts as the domain expert) and various members of the development team be involved in various workshops leading to this formulation. The rationale behind involving some members of the development team is to ensure that there is a general understanding of what is to be achieved and whether this is realistically achievable given the team's skills and expertise. This effort is led jointly by the project manager and the development team lead, who in this case is also the head of AI.

Performance

The lack of confidence in the model built are some of the key reasons why the majority of AI projects never make it to production and are therefore perceived as failure by the organization. Defining appropriate performance metrics is crucial for a successful AI project. As discussed earlier in the "Data Science Development Process" section of Chapter 4, the rationale behind defining performance up front is to ensure that there is a clear alignment between the business and the development team.

Following the workshops between the project manager and the development team, the following was decided:

- The cost of incorrectly saying an application is a good credit risk, formally false positive (FP),[2] outweighs the cost of incorrectly saying an application is a bad credit risk, formally false negative (FN).[3] Stated differently, the system should be designed to minimize false positives. In business terms, the organization would like to minimize the proportion of applicants that later on default.

- In line with the organization's values and principles, fairness/bias and interpretability have been identified as soft performance metrics for the system under development.

---

[2]The system incorrectly indicates that the applicant is a good credit risk when the applicant is in fact a bad credit risk.
[3]The system incorrectly indicates that the applicant is a bad credit risk when the applicant is in fact a good credit risk.

- Although predicting whether an applicant is a good or bad credit risk is convenient, it is often preferable to estimate the probability of the predicted category. Such probabilities provide further insight into the model's level of confidence while allowing more flexibility regarding the interpretation of predictions.

It is important to provide a comprehensive description of the behavior of the system under development to avoid any ambiguity regarding criteria against which the system is evaluated.

Controls

For the sake of simplicity, only a limited number of controls are identified for the performance phase. These focus on defining and documenting the protocol for the evaluation of the system against identified performance metrics. These include, but are not limited to

- The data split strategy and minimum size of data used for training, evaluation, and testing the future model.

- Key characteristics of the data set to be used for computing final performance metrics. For example, sensitive attributes such as age and gender should be in equal (or almost) proportions in the final test data set.

- The system should be able to highlight what predictor variable(s) and value(s) influenced most the model's prediction.

Governance

The development team lead (DTL) in collaboration with the project manager ensures that controls specified at this stage and defined performance are realistic and achievable. Failing to define a reasonable objective for the project is a recipe for failure. In effect, for some convoluted reasons, organizations often have the unhealthy tendency to focus most of their effort on overly complex objectives or projects, which in practice increases the failure rate of AI projects.

Another aspect related to AI projects' failure that is often overlooked when defining the objectives of an AI project is related to the skill set of the development team. Realistic objectives should mean realistic objectives for the development team working on the project. It is the DTL's responsibility to agree to objectives that are realistically achievable by the team, because failing to do so increases both the project's cost and delivery time.

As part of this governance arrangement, the DTL also ensures that the development team has access to the appropriate infrastructure to develop, test, and deploy the model. To achieve this, the DTL liaises with the organization's IT team to ensure that it can provide tools and software to support the development team throughout the project's life cycle. Precisely, the DTL works with the IT department to design and deploy a data processing infrastructure that meets the organization's standards to guarantee user privacy and maintain tight control over who accesses what information within the organization and what it is used for.

As alluded to in the "Soft Performance Metrics" section of Chapter 5, traditional performance metrics such as FN and FP may be conflicting with soft performance metrics in the like of bias, meaning that adjusting the algorithm to make its outcome fair to all groups can sometimes lead to a less accurate model overall. A trade-off between our selected performance metrics and bias may be required, and this needs to be adequately communicated to the business.

Data Collection and/or Interpretation
Controls

Using the right data is essential for the success of any AI project. Controls for this phase typically ensure that the data used for the development of the system meets the expectations set out in the performance phase.

Data for training, evaluating, and testing the credit scoring system is composed of historical credit/loan data collected by the organization over the past five years. This data is provided by the credit risk team in accordance with the organization's data management policies. It includes a wide array of information such as debtors' demographic information and information on how and whether they repaid their loan. The following non-exhaustive list of controls was identified for this phase. One can think of them as a checklist of constraints that the data collection and interpretation phase must meet. Although specific to this project, these controls apply to a wide range of AI projects:

- Adequacy of the data for credit risk assessment
    - As part of the process of assessing the adequacy of the data for credit scoring, key characteristics of the data are unveiled and documented accordingly. Such information helps identify factors that may contribute to fairness/bias-related issues.

- Adequate examination of the data set for bias

  - When the organization has provided a clear definition of what qualifies as a sensitive attribute, one may check for the presence of such attributes in the data set and examine their relationship with the outcome variable. However, when such definition is unavailable, demographic attributes are generally considered sensitive and may constitute a good starting point.

  - Once identified, one needs to measure or assess the impact of these attributes on the outcome and take appropriate actions or steps to mitigate such impact. Stated differently, the approach for mitigating bias needs to be documented along with its limitations if any.

- Compliance of the data with regulations such as the GDPR

  - Data used for credit scoring may often come from diverse sources. It is essential to take necessary steps to ensure that the data is collected and processed within the boundaries of applicable regulations.

- Adequacy of the documentation of data preprocessing and other transformations applied to the data including methodology for handling errors and missing values

- Adequacy of the data labeling methodology

  - In a classification setting like the one at hand, the data needs to be labeled. This labeling is typically done by domain experts who understand the data and the business context well. However, it may sometimes rely on convoluted rules. These along with other assumptions are documented and their adequacy assessed.

- Adequacy of the documentation of the data and understanding of the raw data

  - Variables which are poorly understood, especially by the development team, should not be included in the model's design. This may happen when the data is acquired from a data broker or third party that does not fully disclose how this data is generated. Another problem that often arises is related to the fact that the organization does not properly communicate to the data broker what the data is for (e.g., for confidentiality reasons) and so may end up with some data points that are completely irrelevant to the task at hand.

Governance

The lack of data is arguably one of the biggest challenges for AI projects' failure. As part of the governance arrangement for the data collection and/or interpretation phase, the project manager must ensure that there is a clear and effective strategy in place for the acquisition of the data needed to develop, test, and evaluate the project. To accomplish this, the project manager works with the DTL and relevant data source providers within and/or outside the organization.

Another governance arrangement around data collection and understanding involves defining relevant controls and ensuring that they are adequately implemented and maintained throughout the project's life cycle. The associated role is played by the DTL who delegates responsibilities within the development team accordingly. As part of this process, the development team liaises with the project manager and the credit risk team which should provide adequate documentation of the data.

Often, organizations do not have a clear definition of what constitutes sensitive attributes. When such a definition is unavailable, it is the DTL's responsibility to work with relevant departments within the organization to provide one to the development team.

Model Building
Controls

As with the previous phases, controls for the model building phase are designed to ensure that the model meets expected performance and the latent business objectives. As illustrated below, they can be formulated in the form of questions that need to be answered when building the model.

In the frame of the credit scoring system, the following non-exhaustive list of controls has been identified:

- Are the size of the data set and period it covers adequate for the system under development?

- This information is desirable in the frame of the credit risk assessment project because the confidence in the model's decision generally increases as the size of the training data increases. Additionally, it also informs whether the model's development complies with applicable regulations.

- Lack of training data is often a key contributor to poor performance of ML models. In some circumstances, enough data may not be available for all customer segments, in which case it may be desirable to create different models for various user segments depending on data/features availability. However, poor performance resulting from data availability is more likely to be observed in the early days of the model.

- Is the development team's understanding of the model adequate?

  - Not every member of the development team needs to understand every aspect of the model; however, such understanding is required from the DTL who takes appropriate actions to bring other members of the team up to date.

- Is the model appropriately calibrated?

  - This is desired because we want the estimated probabilities of default to accurately represent the model's confidence level. Stated differently, given a sample of loans, their predicted probabilities of default should be as close as possible to the percentage of default in that sample. Note that for some models, such as logistic regression, the model's estimated class probabilities are already calibrated; however, for some other algorithms such as random forest, these do not reflect the confidence of the algorithm in the prediction and need to be calibrated accordingly.

- Is there a need for the model to be interpretable?

  - It is important to understand where interpretability is needed and what needs to be interpretable.

- Has the model been adequately documented?

  - Building an AI model involves several tasks, such as splitting the data, extracting features that will serve as input to the model, training, and testing the model. These tasks need to be appropriately documented along with assumptions. Such documentation should ensure that the model can be recreated from it.

  - Multiple models may be created during this phase; however, only the best model or the model to be deployed in production needs to be documented.

- What measures have been taken to mitigate bias?

  - Bias may occur at different stages of the development process including the model building phase. When mitigating bias is of interest, one may apply additional transformations to the data if required or rely on algorithms or tools that incorporate bias mitigation properties. This needs to be documented accordingly.

- Is the model and data used for its training versioned and maintained adequately?

  - Maintaining a track record of data used to train an ML model, as well as the training code, is of great importance because it allows not only to switch between different versions of the model and/or data when needed but, more importantly, to keep track of changes that have been applied to the model. Tools such as data version control (DVC)[4] can be used for this purpose.

Governance

Governance arrangements for the model building phase are led by the DTL who defines relevant controls and oversees their implementation. The DTL accomplishes this by collaborating with the project manager to fully understand the behavior expected from the model in a real-life deployment scenario.

Organizations are often reluctant in investing in the right resources for AI projects. However, failing to bring in the right support may not only delay its delivery but hinder its success. This is because implementing certain controls may require additional resources in terms of time, manpower, and training.

---

[4]https://dvc.org

The DTL must liaise with relevant departments within the organization to provide the support the development team needs. Considering this, it is important that adopting the SAIF framework be an organization-wide initiative because implementing controls will generally require involvement from other departments such as the IT, legal, and human resource departments.

Evaluation/Performance Measurement Controls

Controls for the model evaluation phase of the DS process are designed to verify that the model is adequately tested. Below is a non-exhaustive list of controls that have been identified for this phase. They are expressed in the form of questions that must be answered during the evaluation of the model:

- Is the model's evaluation methodology or scheme documented, and performance metrics adequately measured and reported?

    - Documenting the model's evaluation strategy can help identify or rule out flaws in the design of the final AI system.

    - For example, the data used to evaluate the model should reflect the distribution of the training data. As discussed, a technique known as k-fold cross-validation is often used for model validating. When used incorrectly, it may lead to a misleading estimate of the model's performance.

    - It is important to note that the ideal approach for validating a model would be to examine its performance on a large enough test data set that was not used during the model building phase. The adequacy of this data set needs to be examined and documented along with its key characteristics to ensure that it is representative of the training data.

- Have important features in the model been identified and documented?

    - An approach for achieving this consists of altering the model's input data and observing the associated change in resulting predictions. However, certain algorithms can automatically compute feature importance and therefore provide a better understanding of how each feature or variable affects the model's outcome.

- For more complex models, a dependency plot can help illustrate the effect of a feature on the predicted outcome.

- Has the model been assessed by the domain expert?

  - More than often, domain experts or future users of a system would have a better understanding of how to test a model. However, because domain experts are not necessarily expert in AI, their ability to effectively challenge the model depends on how well the model is documented. This points to the importance of the development team communicating with the business to ensure that they understand how the model works. Controls discussed thus far should provide a good understanding of how the model is built and works.

  - Another benefit for involving domain experts is that they can help design effective scenarios for testing the model.

- Is performance correctly interpreted?

  - Although models' performance is often evaluated using metrics discussed in the "AI Performance Metrics Overview" section of Chapter 5, these often need to be interpreted within the business context. In the case of the credit scoring system, the computed credit scores along with associated false negative and false positive are communicated to the credit risk expert who then helps define an acceptable risk level.

Governance

Key roles for the model evaluation phase involve the development team and the domain expert, who in this case is represented by the project manager. The project manager signs off on the system before it enters the deployment phase. The DTL delegates appropriate responsibilities to members of the development team and provides them with necessary support and training if required. Additionally, the DTL works with the business expert to ensure that performance metrics are well understood and interpreted correctly within the business context.

Model Deployment
Controls

Controls for the model deployment phase of the DS process ensure that the model selected for deployment meets the organization's business objectives and values. Below is a non-exhaustive list of controls expressed in the form of questions that must be answered before or during model deployment:

- Is the model interpretable?
  - It is worth noting that depending on the application at hand, one can achieve interpretability through an understanding of the relative contribution of each predictor to the response. Credit risk assessment models traditionally rely on logistic regression, which is relatively easy to interpret. However, other algorithms such as random forest and gradient boosting trees may offer better performance and can automatically compute variable importance which approximates how important the variable is at describing the response. Generally, for more complex models, an outcome-based approach to interpretability can be adopted.
  - It is common to associate a model's interpretability with the ability to explain a single outcome. However, ML models are generally judged through performance metrics, such as accuracy, which reflect their ability to accurately predict the outcome for a relatively large number of observations. This suggests that some effort should be devoted to interpreting performance metrics in the context of a business problem at hand.
- Has the model's outcome been assessed for bias?
  - Examining the model's performance within subgroups of the sensitive attribute can help identify bias in the model's outcome.
  - The appropriate fairness metric must be agreed upon with the business; this is because a model can be "fair" by a metric, but be unfair by a different metric.

- Has the model's management and monitoring strategy been defined?

    - Implementing a model monitoring and management strategy prior to its deployment ensures that errors are proactively identified and dealt with accordingly. As part of this, necessary precautions should be taken to ensure that the model can be easily updated. Model update may be required for many reasons, such as change in business conditions or the model no longer accurately represents the target population for which it is deployed.

    - This also guarantees that an entity within the organization, such as the data science team, owns the system and is responsible for its ongoing maintenance and improvement.

- Is the AI system under development auditable?

    - It is essential to know which version of the code, data, and model is deployed to production, especially when the model is not performing appropriately.

- Is the models' deployment strategy adequate?

    - Typically, approaches for deploying ML models can be grouped into dynamic and static. A dynamic model is often preferred when the data is likely to change over time. For the credit risk example, the system needs to constantly adjust to its target population, and so a dynamic deployment is more appropriate.

- Is the deployed system scalable and secured with appropriate user and permission management?

    - Scalability is not always required, but such a decision usually depends on the number of users of the platform. Collaboration with the business is required to decide on the model's scalability.

Governance

Governance arrangements for the model deployment phase ensure that appropriate controls are defined, implemented, and maintained throughout the model's life cycle. The DTL oversees this phase and works with the project manager along with technical experts from the IT department to ensure that the development team has appropriate support for the deployment.

In some circumstances, the model created by the development team needs to be handed off to a software development team to be reimplemented in a programming language compatible with the organization's software development stack. In such circumstances, the DTL must closely supervise the new implementation and work with the software development team to ensure that it is adequately tested.

Defining appropriate controls for this phase is primarily coordinated by the development team lead who actively collaborates with the IT department and the project manager (also the domain expert in this case) to identify relevant controls. Additionally, the DTL must ensure that the project manager signs off on the system before the deployment.

Conclusion

In this chapter, we discussed the use case of an AI-based credit scoring system to illustrate how organizations can effectively manage the risk around AI through the process – controls – governance model of the SAIF framework. Precisely, we have identified controls for various stages of the DS process and discussed governance arrangements needed both to identify and implement them. Despite a special focus on the credit scoring system, the description of controls and governance arrangements is applicable to most AI projects.

Organizations are increasingly relying on third parties for the development and deployment of their AI capabilities. In such circumstances, an organization should ensure that its AI system has appropriate controls in place before it is used.

Alternative Avenues for Regulating AI Development

AI and Regulations

Thus far, we have discussed the many potential ethical consequences associated with AI systems. More importantly, we presented SAIF, a methodology to help organizations and their stakeholders carry out AI due diligence to prevent, identify, better understand, and mitigate undesirable consequences resulting from the DS practice throughout the development and deployment of AI systems. We have also seen that some countries, mainly developed countries, have introduced regulatory measures such as data protection regulations like GDPR. Such regulations require service providers to explicitly get informed

© Ghislain Landry Tsafack Chetsa 2021
G. L. Tsafack Chetsa, *Towards Sustainable Artificial Intelligence*,
https://doi.org/10.1007/978-1-4842-7214-5_7

consent from their users before collecting their data, including data resulting from their interactions with the service. We argued that, as of today, such regulations are ineffective for multiple reasons including the following:

- Users cannot use some services unless they consent to data collection. Moreover, terms and conditions are generally very convoluted, inaccessible to the user, and often framed so as to allow the service provider to collect more information than needed to deliver the service. In other words, users often find themselves trapped in a system designed to maximize data collection if they want to continue using a service. More generally, the system's framing – that is, how the system is presented to its users and understood by the public – often disguises various operational aspects of the service.

- Regulators are unable to create an environment capable of guaranteeing the effectiveness of their regulatory system. Such an environment would need to involve technological resources, frameworks, tools, and processes for enforcing existing regulations.

- Data protection laws vary widely by country, and some do not have any at all. For developing nations, there are usually other priorities than data privacy. As a result, people in such countries may be left more vulnerable to data exploitation than in countries where there is greater regulatory oversight. It is not unheard of for companies to sell their services in less regulated regions like these solely to capture and monetize data.[1]

It is necessary to understand some of the elements of effective regulation to better understand why a regulation such as the GDPR, despite its potential to create a fair environment for companies and consumers, remains problematic. The Oxford Learner's Dictionaries[2] defines the word regulation as "an official rule made by a government or some other authorities." Typically, a regulation involves three main components: legislation, enforcement, and adjudication (Swire 1997). The respective legislation defines appropriate rules, whereas enforcement involves initiating punitive actions against people or organizations that violate the legislation. The third component, adjudication, is the process of deciding whether a

---

[1] https://privacyinternational.org/long-read/3390/2020-crucial-year-fight-data-protection-africa
[2] www.oxfordlearnersdictionaries.com (accessed 28/11/20)

violation has taken place and imposing sanctions accordingly. While the first two components are relatively easy to implement in the context of AI, adjudication is not at the time of the writing of this book.

Consequently, for an organization to comply with regulations such as GDPR is one thing. However, meeting the latent expectations of such regulation is a completely different matter and requires a willingness to invest the required effort voluntarily. Simply complying with regulations does not guarantee ethical behavior. Realizing this requires an understanding of the difference between "compliance" and "ethical behavior." Compliance is simply acting in accordance with requirements (usually imposed upon by a regulatory body) so as to gain reward or avoid punishment. Ethical behavior, instead, requires adopting pro-social behavior voluntarily and is motivated by the desire to do what is right. This effectively requires self-regulation.

The distinction between ethical behavior and compliance suggests that ethics and compliance are both desired and aim to achieve the same goal. To achieve such a goal, compliance will ensure that all regulatory and legal requirements are met. Ethics ensures that the organization continues to behave in a respectable manner toward the organization's stakeholders, when no one is watching. However, organizations find it easier to comply with existing regulations since it removes the stage of figuring out how to behave that is required when thinking about what one wants to do to behave ethically. This probably explains why compliance is often preferred over ethical behavior, which in some cases may be subjective. Consequently, shared objectives across organizations, set outside the scope of individual organizations, are needed to establish a conduct that benefits not only the organization but also its stakeholders and the society. Specifically, such shared objectives prescribe the minimum requirements for ethical behavior, in respect of the development and deployment of AI systems. This set of requirements must be followed by any organization that develops and/or sells AI products.

This chapter examines various approaches for establishing such objectives. More precisely, it investigates various mechanisms for regulating the development and deployment of AI systems. The remainder of this chapter is structured as follows:

- First, we discuss self-regulation as a mechanism to instill ethical behavior and highlight some of its limitations.

- We then examine the idea of a third-party regulatory market as another mechanism for instilling ethical behavior throughout the development and deployment of AI systems.

- Next, mechanisms for a fairer redistribution of revenue related to data monetization are investigated.

- Finally, we make some concluding remarks.

Self-Regulation and the Practice of DS

There is no single definition of the term "self-regulation." On the one hand, "industry self-regulation," as is often referred to in an industry context, implies that the industry rather than the government is doing the regulating. In certain circumstances, self-regulation can still happen even if the industry is only involved in one or two of the three components of regulation discussed earlier (legislation, enforcement, and adjudication). For example, the government may define rules and objectives for the industry or grant certain organizations legal power to sanction violators (Gupta and Lad 1983). In this scenario, the industry would be responsible for adjudication. On the other hand, from an individual's perspective, self-regulation can be thought of as the ability to align one's behavior to rules prescribed within one's environment. Within an organization, self-regulation at the professional level can be associated with behaving in line with the general code of conduct (organizational and societal ethics). Consequently, because individuals already self-regulate to the society, as long as a company's ethical beliefs are aligned with those of the broader society, self-regulation is more likely to produce the desired ethical effect. In line with this, Andrew (Andrew 1998) argues that self-regulating approaches for instilling ethical behavior at the professional level are more successful than policy-like measures because the former builds a culture of trust within which ethical behavior is regulated through the members of the society that share this culture. However, most organizations are profit driven and are unlikely to take any action that may affect their profit unless such actions are either imposed by a regulatory body or demanded by their customers.

While self-regulation at the individual level provides a framework to help mitigate ethical consequences resulting from the practice of DS, it does not necessarily produce the desired effect at the industry level, especially when the industry takes over all three aspects of regulation. In the context of AI, this can be attributed to the conflicting nature of the relationship between the data object and the industry's desire to create revenues and drive innovation. Additionally, AI research and development is in constant evolution and change. Organizations continue to seek innovation and insights through new use cases and sophisticated AI models as the competition for customers' attention becomes more and more difficult. Such a race to innovation and competitive advantage results in a lack of alignment within the industry, making industry self-regulation difficult and often undesirable by the members of the industry itself, especially if they must share data. It is therefore not surprising that some of the largest users of DS, including Facebook, Amazon, and Google, have announced intent to self-regulate their services. However, this is unlikely to solve all the problems we are facing today. In other words, effective regulation is likely to be hindered by the industry's self-interest.

Another deterrent for industry self-regulation in the context of AI is the potential to promote cartel-like conduct: industry self-regulatory arrangements are designed by a few powerful organizations to advance or promote narrow interests to the detriment of smaller companies and the wider community. At the time of the writing of this book, the development and deployment of AI systems is dominated by a handful of organizations, in part because of their overwhelming presence and influence over the Internet. Consequently, it is common sense to assume that under an industry self-regulation regime, rules are likely to be designed to maintain the competitive as well as the financial advantage of these corporations. This is not specific to AI: a recent lawsuit against the tech industry giant Apple on its App Store policy regarding in-app purchase options is a perfect illustration that organizations with power are likely to create asymmetric rules that disfavor consumers[3] and competing companies. To provide some context around the lawsuit, Epic Games, the company behind the lawsuit, was kicked off the App Store because it introduced a direct payment system to its popular game *Fortnite* to bypass Apple's 30% fee. It is noteworthy that app developers are prevented from using the App Store unless they agree to Apple's exorbitant 30% fee. Following the lawsuit, Apple announced it would drop its 30% cut of in-app purchase down to 15%.[4] Like the Epic Games' lawsuit, antitrust charges have been brought up against other powerful organizations such as Google[5] and Amazon.[6]

The attitude of these organizations toward competition can be summarized in the following quote from the "Investigation of Competition in Digital Markets: Majority Staff Report and Recommendations" report by the Democratic Majority of the Subcommittee on Antitrust, Commercial and Administrative Law of the Committee on the Judiciary of the US House of Representatives:

Companies that once were scrappy, underdog startups that challenged the status quo have become the kinds of monopolies we last saw in the era of oil barons and railroad tycoons. Although these firms have delivered clear benefits to society, the dominance of Amazon, Apple, Facebook, and Google has come at a price. These firms typically run the marketplace while also competing in it – a position that enables them to write one set of rules for others, while they play by another, or to engage in a form of their own private quasi regulation that is unaccountable to anyone but themselves.[7]

---

[3]The extra cost on the product is often passed on to the consumer.
[4]www.theguardian.com/technology/2020/nov/18/apple-to-reduce-its-cut-from-in-app-purchases-as-it-faces-new-lawsuit-from-fortnite-maker
[5]www.bbc.co.uk/news/business-54619148
[6]https://news.sky.com/story/amazon-hit-with-antitrust-charges-by-eu-regulators-12129148
[7]https://judiciary.house.gov/uploadedfiles/competition_in_digital_markets.pdf

The above quote suggests that organisations that have accumulated power set rules to their advantage. The distribution of power within the AI industry is not so different, because the AI industry is currently steered by a handful of powerful organisations that are also at the forefront of AI research and development. Therefore, it can be argued that the AI industry may not be ready for self-regulation.

Despite the potential challenges that industry self-regulation presents in the context of AI, there are also some benefits to self-regulation within the AI industry. Importantly, the industry's knowledge of the subject matter is often cited as one of the biggest advantages of industry self-regulation over government regulation. This knowledge advantage is critical to the effective regulation of the increasingly complex and multidisciplinary nature of AI technology. Still from a technical perspective, AI systems are often described as opaque because their inner working is complex, poorly understood, and nearly impossible to retrace by humans after the facts.[8] Under such circumstances, the collective industry is probably better equipped to do the regulation than a government agency (Swire 1997). In addition, the fast-paced innovation in AI development and deployment creates an extra level of complexity, which further limits the ability of any government agency to keep regulation up to date.

Another benefit of industry self-regulation in the context of AI is that it could set minimum standards for the development and deployment of AI technology on the global basis. Doing so, organizations can be held accountable by their industry peers for their actions irrespective of what is allowed by national jurisdictions. This is particularly important, especially for developing countries. In such countries, western corporations continue to unscrupulously harvest data, often unbeknown to the population, for their own interest (Hendricks, Mads, and Silas 2018) because of the lack of data protection regulations. Moreover, even where regulations exist, the local context may still facilitate undesirable or unethical behaviors. This is unfortunate because certain organizations may not hesitate to take full advantage of such circumstances. In light of this, it can be argued that industry-level regulation may deter organizations from misusing the AI technology where rules and regulations are either nonexistent or not enforced.

As the competition for customers' attention through personalized services[9] continues to intensify, the development of AI systems remains under what can be described as proprietary opacity. This refers to the fact that organizations are unwilling, despite public discourse, to disclose the inner working of their AI systems for competitive reasons. It can be argued that under these circumstances industry self-regulation provides greater incentives for compliance because it becomes an organization's responsibility to demonstrate that it is operating by the industry's standards.

---

[8]Meaning that once an AI system has made a decision, it is often difficult if not impossible to describe with certainty the steps that led to that decision.

[9]These are typically performed using AI and related technologies such as IoT devices.

Other challenges that industry self-regulation can help address are related to how AI development happens. Beyond opacity, AI development is often

- Discreet, that is, it can be developed with limited visible infrastructure

- Diffuse, meaning developers of a single AI component may be in different regions

- Discrete, that is, conscious coordination is not required for the development of separate components of an AI system (Scherer 2015)

With the industry regulating itself, a minimum set of standards can be defined for every participating member to comply with.

In this section, we have presented arguments both in favor and against industry self-regulation in the context of AI. While industry self-regulation presents some benefits, it emerges from the above discussion that it is insufficient when applied alone and requires government involvement/intervention in some shape or form for its effectiveness. The next section examines a mechanism through which the government can intervene to enable effective industry self-regulation.

Toward a Third-Party Regulatory Market

In the previous section, we argued that, while seemingly undesirable in the context of AI, industry self-regulation exhibits interesting properties under certain circumstances. In this section, we examine a regulatory arrangement through which the desired positive behavior within the AI industry can be designed and achieved for the greater benefit of the community as a whole. Regulation involves three key components as highlighted earlier in the "Self-Regulation and the Practice of DS" section: legislation, enforcement, and adjudication. Typically, self-regulation implies that all or some of these components are being performed by the industry itself. However, the government may instead maintain control over one or two of the three components of regulation. This can be achieved through government empowering government agencies to perform the corresponding aspects of regulation. In this section, we will now examine the roles that the main actors of the regulatory arrangement must play to guarantee continuous and sustainable development and innovation through AI while efficiently managing and mitigating the associated risk to the society. These actors are the government, technology companies/the AI industry, and customers.[10]

---

[10]Customers, as alluded to earlier, are users of the products and services developed by technology companies operating within the AI industry.

Roles in the Regulatory Arrangement

To better understand the roles of these three actors with respect to regulation, we first examine a remarkably interesting aspect of the GDPR: At the time of writing, a user has the right to be "forgotten" under the GDPR, meaning that all of their data is destroyed. However, the user, after invoking its right to be forgotten, has no ability to verify that this actually took place. Ironically, neither does the regulator, meaning that the best option is to take the service provider's word and hope it actually removed relevant records from its systems. In the case of GDPR, the legislation and enforcement elements of regulation have been defined by the government or law makers. However, as this example shows, an effective adjudication mechanism is both lacking and yet to be defined.

At the center of various discussions around ethical development, deployment, and use of AI sits the matter of how to design an effective third-party regulatory structure.[11] Acting as a middle entity between government actors and the AI industry, such a regulatory structure should incentivize organizations to design innovative regulatory systems that can keep up with the complexity of AI technologies. Specifically, the government should empower third-party organizations to collectively take over the adjudication element of regulation. The main benefit of such a regulatory structure is that it creates an environment where nonprofit and for-profit private or government agencies can compete for the right to develop tools and technologies needed to support regulations around advances of AI technologies. We argue that such a regulatory structure provides a mechanism through which effective regulation can be achieved. We refer to this structure as a market because it allows organizations to acquire certain rights. For example, with a third-party regulatory market in place, organizations wanting to operate within specific industry sectors will require an accreditation or a certificate from members of the regulatory market, and by the same means make their technology available to the members of the regulatory market for regular audit. The widespread adoption of AI across industries, not all of which conform to the same rules, suggests that the regulatory market may need to design industry-specific certificates depending on the level of safeguarding required.

The idea of a regulatory market through which organizations acquire the right to operate within a given industry (by obtaining a certificate from accredited groups or organizations) is relatively well established in industries such as healthcare. For example, the Health Insurance Portability and Accountability Act of 1996 (HIPAA) is a US privacy law to protect medical information. While not being required to certify their compliance, organizations operating in the healthcare industry in the United States generally complete a HIPAA certificate to demonstrate that they meet the standards of the privacy,

---

[11]Led by Professor Gillian Hadfield from the University of Toronto.

security, and breach notification rules of HIPAA as defined by the Health and Human Service (HHS). In this setting, the HHS being a government agency provides what can be perceived here as the legislation, that is, the HIPAA legislation, whereas third-party HIPAA training companies provide organizations wanting to operate in the healthcare industry with a certificate demonstrating that they adequately comply with the HIPAA legislation.

The HIPAA example illustrates how a regulatory regime, with a third-party market or industry responsible for auditing organizations to ensure that they meet AI standards for their specific industry, could work in practice. Fields of property and building consultancy constitute other industries where individuals acquire the "right" to operate or provide their services by obtaining a certificate from an accredited body. Specifically, chartered surveyors must undertake a degree accredited by Royal Institution of Chartered Surveyors (RICS) and pass an assessment of professional competence (Ford 2007).

Despite its attractiveness and vast potential, establishing a third-party regulatory market is challenged by the fact that major research and development in AI is carried out by only a limited number of organizations. This means that, realistically, only a handful of companies could currently effectively compete for a spot in the regulatory market. The AI industry is well aware of these limitations, including the knowledge asymmetry within the AI community in general. Limited availability of data is a key contributor to this knowledge asymmetry, meaning that organizations that have "the data" are in a better position to conduct AI research and consequently innovate than organizations that do not have access to such data. This is corroborated by the ongoing campaign by organizations such as the European Data Portal[12] and the University of California, Irvine (UCI) to create publicly and freely available data repositories.

Another challenge that a third-party regulatory market for AI may face is related to the matters of rules of admission and membership of the market. Such rules must be designed by the government in collaboration with the industry, not-for-profit organizations, and a diversity of users' rights advocates. The main issue to address is that the policy for the admission to the regulatory market may introduce the risk of reverting into a cartel-like behavior often observed within a self-regulatory regime. For example, if the regulatory market becomes both dominant and profitable, its members may be tempted to prevent new entrants that are attracted by its success from joining the market to maintain their monopoly profits.

A diverse range of organizations of various sizes from small to medium enterprises is needed to establish and maintain an environment that does not favor the interests of a smaller group. To ensure such diversity, specialized training programs may be required to facilitate the access of smaller and

---

[12]https://data.europa.eu/en

cash-restrained companies to the regulatory market. This may help to ensure that organizations with limited ability to conduct research can acquire the knowledge needed to participate in the regulatory market.

The Role of Customers in the Regulatory Arrangement

Thus far in this chapter, we have largely focused the discussion on actors other than the customers. However, it is needless to say that the customer plays and must continue to play a fundamental role by influencing the design of AI-related legislations and demanding that organizations operate by the highest standards. This is necessary not only because AI systems are increasingly becoming pervasive in almost every aspect of daily life but more importantly because

- Reference benchmarks for the quality of AI systems are lacking.

- Fundamental aspects of the quality of AI systems, such as accuracy, and adherence with data privacy laws cannot be observed.

- Increasing demand for regulation will stimulate the rise of private for-profit and not-for-profit regulatory firms. While this may seem light years away to some readers, the industry is certainly moving in the right direction as some organizations have expanded their service offerings to include AI and/or data audit in some way or form.[13]

For the above to work in practice, customers must continue to exercise their ability to influence policies and how organizations operate. Precisely by demanding high standard services, customers can selectively choose the service provider which best meets their expectations. In doing so they not only encourage competition but, more importantly, they implicitly dictate standards that organizations or institutions should follow to remain relevant.

Let us illustrate this through two examples.

WhatsApp messenger (or simply WhatsApp) is a mass messaging platform owned by Facebook. It is without doubt the most popular mobile messaging application with over 1.5 billion users worldwide as of 2019. WhatsApp recently announced that WhatsApp users must agree to its new terms of

---

[13]For example, some organizations provide AI and data audit services to venture capitalists wanting to invest in fast-growing technological start-ups that rely on AI for the services they offer.

service and privacy policy by February 8, 2021, if they wanted to continue using the service. Following this, WhatsApp faced a massive backlash, shown by the fact that millions of its users switched to competing services Signal[14] and Telegram.[15] Such a backlash was primarily motivated by the lack of privacy and the level of access to users' personal information that Facebook and its subsidiaries will gain once a user has agreed to the new terms of service and privacy policy. In response to users' concerns and criticisms, and further to control the damages, WhatsApp updated its website to introduce an FAQ section through which it attempted to clarify that the new privacy policy was only affecting businesses using WhatsApp for customer service purposes (WhatsApp LLC 2021). Specifically, the company states

> We want to be clear that the policy update does not affect the privacy of your messages with friends or family in any way. The changes are related to optional business features on WhatsApp, and provides further transparency about how we collect and use data.

—FAQ, WhatsApp LLC (WhatsApp LLC 2021)

In this attempt to reassure its users, the company further clarifies that it cannot see personal messages or hear users' calls or see users' shared location, and neither can the parent company Facebook.

Another action that the company undertook in response to user complaints about the new terms consisted of delaying the effectiveness of the terms to May 15, 2021. Nevertheless, users looking to continue using the service will have to accept the terms and conditions by this date.

Considering WhatsApp's response to the incident, some would argue that this was simply a misunderstanding resulting from the poorly drafted terms of privacy that confused some users and alienated those who switched to competing services. Some would however argue that this was a genuine attempt of the company to use its dominance of a developing market at the expense of its users. In any case, the lesson learned from this incident is that users are willing and can have enough power to influence and, in some cases, affect the behavior of corporations, especially now that the competition for users' attention is fiercer than ever.

Another example reminiscent of the users' ability and willingness to influence not only corporate but also government actions that do not meet their expectations is provided by the controversy surrounding the UK's general certificate of secondary education (GCSE) and A Level grading during the COVID-19 pandemic.

---

[14]https://signal.org/en/
[15]https://telegram.org

In 2020, the COVID-19 pandemic forced the cancellation of all secondary education examinations in the UK. Consequently, the government decided to use an algorithm to standardize teachers' predicted grades to grade students for the year 2020. A grading standardization algorithm was then designed and implemented by the accredited regulatory bodies in nations within the UK. For example, in England and Scotland this algorithm was created by the Office of Qualifications and Examinations Regulation (Ofqual) and the Scottish Qualifications Authority, respectively. Grades issued by the algorithm were met with public outcry. Specifically, it was observed that the algorithm advantages students from higher socioeconomic backgrounds at the expense of those coming from lower socioeconomic backgrounds, a population that generally attends state schools. In response to the outcry, the government decided to revisit its position by either returning to grades predicted by teachers or selecting whichever grade is the highest between the grade predicted by the grade standardization algorithm and the teachers' predicted grades.

Similar to the WhatsApp example, this illustrates that users have enough power to influence and affect policies. But one question remains: "Will they do it?" One factor in favor of organizations is that to impact large corporations, the number of users demanding change or better policies must be in the order of millions. Nevertheless, with general awareness around data misused or poor data-related decision-making through AI on the rise, users are likely to drive the desired regulations. For example, according to a survey conducted by Pew Research Center in 2019, 81% of Americans think the potential risks of data collection by companies about them outweigh the benefits. Consequently, it is reasonable to believe that users will continue to push for more regulations by changing the government's attitude toward the issue (Brooke et al. 2019).

Toward a Fair Distribution of Data-Related Revenue

Data is arguably inherent to the development of AI systems. This data in most if not all commercial settings of AI can be argued to be a characteristic attribute of the individual from whom it is collected. Such a relationship creates a one-to-one association between a datum and the individual from whom it is collected. Consequently, as discussed in the "Data – The Business Asset" section of Chapter 2, the individual is often turned into a commodity. To illustrate this, let us use the very primitive example of the analogy of the relationship between the milk farmer, a milk cow, and the cow milk.[16] The relationship between the farmer, the milk cow, and the cow milk (hereafter the milk-cow relationship) is in essence similar to the relationship between

---

[16]This analogy is particularly interesting given the many campaigns against farming techniques.

the data controller, the data subject, and the data (hereafter the controller-data subject relationship) except that the data subject is of course not a commodity.

Another difference between the milk-cow relationship and the controller-data subject relationship is that the farmer often must invest some of the revenue from the milk sale into the cow. Meaning that the cow in the right circumstances is a direct beneficiary of its milk production.[17] Depending on an organization's commercial strategy, this is not always the case when it comes to today's data-driven industry – the data subject does not always directly benefit from the processing of its own data. It is probably simpler to look at this through the type of services an organization provides. Typically, we can distinguish two categories of services: paid services and free services or commonly referred to as freemiums. Free services as the name suggests are services offered by an organization to its users free of charge. However, it can be argued that users receive the service free of charge only in exchange for their data. Such services, among others, include social media platforms and free search engines. Therefore, in this setting, customers are commonly referred to as users instead. In a paid service setting, as opposed to a free service setting, the customers through their payment actively offset the operating cost of the organization offering the service they receive. However, still within the context of paid services, the data resulting from the interaction between the service provider and the customer remains the sole property of the service provider. Stated differently, the customer does not only contribute to the creation of the data but also supports the data controller's revenue by paying for the service. However, the customer arguably benefits from revenues generated by the service provider from data monetization. Such asymmetric revenue sharing models are a practice that many may disapprove of. In this section, we discuss alternatives for a fairer distribution of data revenues.

Data Tax

Recent years have experienced an increasing number of calls for data taxation and fair redistribution of data dividend. It can be argued that data taxation may force organizations to adjust their behavior accordingly and therefore reduce "exploitative" behavior that most modern organizations seem to exhibit nowadays. Conceptually, one can think of a data tax as a tax somewhat like the carbon tax, which in its simplest form is a fee imposed on the burning of carbon-based fuels (coal, oil, gas). At its core, the carbon tax is intended to make organizations pay for their contribution to atmospheric pollution, as a result of its effect on global warming, and therefore strongly incentivize the use of carbon-based fuels in favor of other, cleaner sources of energy.

---

[17]Research shows that the cow's comfort has a positive impact on milk quality (Krawczel and Grant 2009).

Similar to the carbon tax, a tax on data would penalize organizations proportionally to the volume of user data[18] they collect to build or develop their business. The proportional structure of this tax will ensure that smaller organizations such as start-ups can still compete or enter the market in the marketplace. For example, organizations may collect data up to a certain volume free of charge. Although technological companies are often on the front page on matters or discussions regarding data monetization, it is important to emphasize that they are not the only organizations that collect, develop, and create insights from data to increase profit. This practice is well established in virtually every industry sector where there are profits to be made such as the banking industry, insurance industry, and the pharmaceutical and healthcare industries. The main benefit of the data tax is that it is probably the simplest and most effective way to guarantee a more equitable distribution, at least to some extent, of the unprecedented profit organizations continue to generate through data. There are multiple arrangements through which such redistribution can be achieved. For example, the resulting tax could be directly redistributed among citizens in the form of a payout once a year. Another alternative may consist of investing in other public services such as healthcare, education, and/or public transport.

A challenging aspect related to the practical implementation of a data tax is the lack of operational transparency about what data is being collected and what it is used for. This lack of transparency limits the ability to effectively determine how much data an organization has collected from its users and what it used that data for. Nevertheless, there is enough evidence to suggest that organizations generate significant profit from this data.

The introduction of the GDPR regulation has provided valuable insights into what can be perceived as initial building blocks for a practical implementation of a tax on data. This can be illustrated through the example of cookies. For some context, cookies are small files that a website keeps on a user's device and then uses it to monitor the user and remember certain information about that specific user. More informally, cookies are information saved about the user, and they track that specific user as they browse the Internet. Typically, there are two types of cookies, first-party cookies that are placed by the sites that the user visits and third-party cookies generally placed by online advertisers to better tract users' interest and serve them adverts accordingly. While cookies can often be used to improve online user experience, their real value resides in the profit an organization can generate by monetizing it. The enforcement of the GDPR regulation has resulted in organizations having to inform and request explicit consent from users for each type of cookies they collect. As a result, at the time of this writing, organizations under the GDPR routinely ask users to select which cookies they are happy for an organization

---

[18]This data is not limited to personal data, but also includes data resulting from the interaction of their users with their services.

to set on their devices. Typically, as seen in Figure 7-1, users are asked to choose whether they would like performance, functional, and targeting cookies to be set on top of the so-called "strictly necessary cookies." Whether the strictly necessary cookies are indeed necessary is a different discussion. Nevertheless, this layered structure provides a basic mechanism that could be relied upon to

- Monitor and track how much data an organization collects from its users

- Implement a multitier tax strategy where each organization receives an allowance consisting of a fixed volume of data it can collect beyond the strictly necessary cookies without incurring any data tax

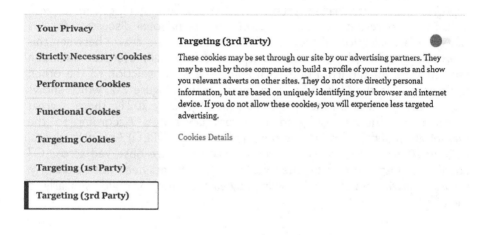

Figure 7-1. *Illustration of various categories of cookies*

For the above arrangement to work in practice, new regulations would be needed. For example, under such regulations, organizations would be required not only to label the data they collect in accordance with the above discussed categories of cookies but also to report the volume of data collected to the appropriate regulatory body which in order to limit abuse must also set appropriate rules for what cookies are deemed strictly necessary on a per-industry basis. Setting such rules is however a tedious process that is highly likely to result in an incomplete set of rules. Another alternative consists of giving organizations a quota of strictly necessary cookies they can collect tax free.

From a technical perspective, for-profit and not-for-profit organizations may compete for the development of software and/or standards to help regulators achieve the above. For example, an Internet browser may integrate plugins that users can install to monitor and report to the relevant regulatory body the volumes of data that individual organizations collect from users' devices.

Another concern often raised against the idea of a data tax is the absence of a consensus around a universally accepted valuation method for data. It can be argued that a cross-border valuation of data is only required when data is being moved from a data regulatory zone to another. A practice that is either prohibited by law or discouraged by most data protection regulations. In this vein, governments or regulators may decide, somewhat arbitrarily, on a quantifiable data valuation that works for them. The establishment of this internal[19] valuation of data should be the priority for decision makers as the approach guarantees that governments, generally developing countries, with little to no representation and/or power in international discussion forums can get a fair valuation of their data. Similarities can be drawn between this approach and the oil price which despite its universally accepted valuation remains very volatile from one economic zone or jurisdiction to the other. For example, in the United States alone as seen in Figure 7-2, which shows state taxes and fees on motor gasoline (as of January 2021), there is great variability on the tax on gasoline across US states. According to the Organization of the Petroleum Exporting Countries (OPEC) Annual Statistical Bulletin 2019, similar variabilities on the tax on oil are observed across G7 countries. This corroborates the idea that a tax on data need not be uniform across countries but should be defined by each country or region similarly to the tax on oil.

---

[19]Internal because it will be specific to a regulatory zone or country.

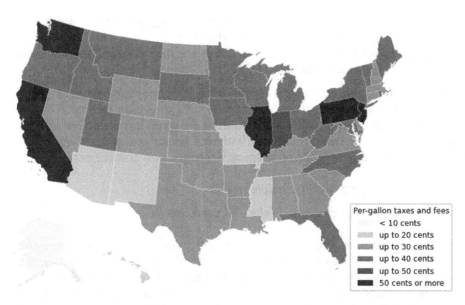

Figure 7-2. Illustration of the interstate variations in the price of one gallon of gasoline across the United States as of January 2021. Source: US Energy Information Administration, Petroleum Marketing Monthly

Data Sharing

We stand at a crossroad between a Web "for everyone" – one that enables all people around the world to improve their life chances and reduces inequalities both between and within countries – and a "winner takes all" Web that further concentrates wealth and political power in the hands of a few.

—The World Wide Web Foundation, 2014–2015 Index

The above quote from the 2014–2015 World Wide Web Foundation (WWWF) annual report on the contribution of the Web to social, economic, and political progress was a clear acknowledgment of the direction that the Internet had taken at the time of the release of the WWWF 2014–2015 Index report. Unfortunately, with the increasing reliance on AI and related technologies, the situation has gone from bad back in 2015 to worse in 2021.

It is argued that data sharing could help reduce inequalities and diversify economic and political power. Precisely, data sharing is understood as providing third-party organizations access to otherwise "proprietary" data sets to generate value. While this approach has seen relative success, it is however unrealistic in a commercial setting to expect organizations to provide competing organizations with the access to the data they collect and store.

This is corroborated by the fact that data sharing agreements are simply commercial agreements in the sense that organizations sharing data are compensated for their effort. In this process, the data creators which in most cases are the users are cast aside.

A fair and equitable alternative to the current data sharing practice is one where users could be compensated for the data they have created. This can be achieved through a data marketplace where users can selectively decide who they want to share their data with and be compensated in exchange. Practically, this can be achieved seemingly through two components:

- An online data marketplace
- A plugin[20] that sits on the end user's device

In this configuration, the plugin with the user's authorization would create a duplicate of all cookies that are collected from the user's device along with a reference to the company that placed the cookie. It would then anonymize duplicated cookies accordingly and send them over a secure connection to the marketplace. Because one person's data is worth little to nothing, the marketplace would work slightly differently from other marketplaces in the sense that one would not be able to buy information/cookies related to a single individual. Instead, the marketplace would allow organizations to buy data of which the underlying structure meets certain criteria. For example, an organization interested in online purchase preferences within a given geographic region may search and buy anonymized cookies from users within the specified region who either searched for or purchased a product online over let's say the past 24 hours. Users meeting those criteria then receive a percentage of the proceeds every time their data is sold.

The above-described marketplace is in essence similar to any other data marketplace; however, it differs in that it allows users to be compensated for their data.

Conclusion

Developing an AI strategy has become a priority for organizations such as technological companies, large retails, banks, and insurance providers. However, realizing and/or maximizing the benefits of such strategy often leads to exploitative practices and behaviors that most people would disapprove of. Regulations constitute a typical mechanism for protecting consumers. Unfortunately, AI, currently lacks such mechanisms of control or effective regulations because of its complexity, rapid development and innovation.

---

[20]Also referred to as add-in or add-on, a plugin is a software component that adds a specific feature to an existing computer program.

This chapter discussed various approaches for regulating AI and establishing an AI ecosystem that is beneficial for every member of the society. Precisely, it examines government regulation and self-regulation regimes for the development and use of AI along with associated services. Because of the complementary nature of both types of regulatory mechanisms, it is argued that a regulatory arrangement that combines the two mechanisms is best suited for effective regulation of AI.

Also examined are mechanisms for a better redistribution of the growing profits produced by users' data along with related services. Precisely, data tax and data sharing are highlighted as effective approaches to make sure that everyone has a share of the wealth created from data.

AI in the Medical Decision Context

AI Impact on Clinical Decision-Making

The straight line, a respectable optical illusion which ruins many a man.

—Victor Hugo, *Les Misérables*

This famous quote from the 19th-century French poet, novelist, and dramatist Victor Hugo somewhat summarizes the complexity of any decision-making process. Typically, a straight line defines the shortest path between two points or objects. These objects need not be physical objects and may simply be an abstract concept, such as an objective or a resolution, in which case the straight line is assimilated to the shortest process one would take to achieve such objective. Through the straight-line metaphor, Victor Hugo attempts to establish the necessity of careful examination of the implications of every decision to be made. Stated differently, it is in the decision maker's best interest to understand, probably to the best of their abilities, the implications or consequences that may happen as a result of the decision being made.

Following Hugo's logic, the decision maker should aim to make a decision that results in the most desired or the least harmful set of consequences.

© Ghislain Landry Tsafack Chetsa 2021
G. L. Tsafack Chetsa, *Towards Sustainable Artificial Intelligence*,
https://doi.org/10.1007/978-1-4842-7214-5_8

For example, almost everyone has experienced a day where one is about to leave the warm of the house and suddenly it looks like it is about to rain. A typical dilemma in such circumstances revolves around deciding whether to take an umbrella or not. The corresponding decision generally takes into consideration factors such as the personal assessment of the likelihood of the rain falling, and how much one values not getting rained on, on that specific day and time. For the sake of this example, on the one hand, we assume that the person wanting to choose between the two options (taking an umbrella and not taking an umbrella) is going to a job interview. Most people would likely take an umbrella, because showing up at an interview wet is not the kind of first impression they want to give to their interviewer. On the other hand, if going to walk the dog, one probably would not mind that much.

We will refer to this example throughout the remainder of this chapter as the rain example.

As illustrated through the rain example, choices are generally not inconsequential; therefore, the decision maker is often expected, to the best of their ability, to evaluate available alternatives and then choose the most convenient option given the circumstances. Central to this decision-making process is the ability of the decision maker to assess the likelihood of the events relevant to the decision maker's choices. In the rain example, this would be the likelihood of rain. This approach to decision-making can be generalized under the subjective expected utility theory resulting from the early work of John von Neumann, Oskar Morgenstern, and Leonard Savage. Formally, the subjective expected utility is defined as an approach to decision-making under risk that allows for subjective evaluation of both the variable under evaluation and the probability associated with them (Shanteau and Pingenot 2009). Stated differently, one can think of the subjective expected utility as a framework through which the decision maker structures the decision. This theory involves three key concepts:

- Decision-making under risk or uncertainty (e.g., risking getting rained on and showing up wet at the interview)

- Utility or value expressed in terms of gain associated with each of the available alternatives (e.g., in overly simplistic terms, whether we mind getting rained on)

- Probability of each of the alternatives happening. For instance, our individual assessment of the likelihood that it is going to rain versus the likelihood that it is not going to rain

Note that we generally do not express such probabilities using numbers as in economic textbooks. Instead, it is common to rely on a relative comparison/ ranking of available alternatives. We rely on our own experience, knowledge, and the context in which we are to establish such comparative analysis. For example, one may think that it is going to rain because it has been raining on and off all morning long. This is generally interpreted as a "guess" not only because of our limited ability to process the information we have at our disposal but also because we cannot quantify neither how "right" or "wrong" we are. Nonetheless, such a guess is often good enough for most people. Put simply, under the right circumstances, we are willing to settle for an option even when we may be fully aware that it is not the optimal option, where the optimal option is the one that maximizes the expected utility.

This approach toward decision-making, consisting of settling with a less-than-optimal option/alternative, is well formulated by the concept of bounded rationality pioneered by the cognitive scientist Herbert Simon (Simon 1957). The idea behind the concept of bounded rationality is that, because people have limited processing ability, they may be happy with picking an option that is just good enough even though it may produce errors in some cases. Considering the rain example, we obviously can make a more informed decision (probably less prone to error) if we know the global sea surface temperature, sea surface level, and how to use that information to assess the likelihood of the rain falling at any given point and time in the region where we are.

A direct consequence of our limited ability not only to gather information but more importantly to efficiently process it is that we tend to rely upon or seek other alternative approaches to help assess available options. Considering the rain example, most people routinely rely upon the weather forecast services to inform their decision on whether it is going to rain and therefore whether to take an umbrella when going outside to an interview on a given day or to go on a getaway weekend. Precisely, one would use the weather forecast service's prediction or estimated probability of rain to decide between taking an umbrella and not taking an umbrella.[1] Services such as the weather forecast service are built on top of sophisticated mathematical models,[2] requiring enormous amounts of data and computing power to estimate the true probability of an event happening. Although the resulting predictions may still be erroneous, they are generally good enough to be useful. The example of the weather forecast service is not isolated; significant effort has been devoted to the development of heuristics or strategies to help and/or support decision-making in various fields.

---

[1] Obviously, we do not deal with probabilities directly, because the weather forecast service for convenience generally converts the probability into a prediction.
[2] www.ecmwf.int

This chapter discusses AI and decision-making in the context of healthcare. Precisely, we examine the relevance of AI in medical decision-making and investigate its limitations as of today.

The remainder of this chapter is organized as follows:

- First, we discuss the relevance of AI in medical decision-making.

- We then illustrate the use of quantitative predictions to support decision-making in healthcare.

- Next, we examine some of the challenges of AI-assisted decision-making.

- Finally, we make some concluding remarks.

The Need for AI in Medical Decision-Making

According to the bounded rationality principle, decision-making is generally suboptimal, meaning the decision maker chooses an option that satisfies a certain acceptance criterion as opposed to the objectively best option (Campitelli and Gobet 2010), which may in some circumstances be impossible to reach. Such approach to decision-making is often observed in medical decision-making where clinicians must routinely choose between alternative options given limited information. Without loss of generality, they may navigate through these alternatives by attempting to maximize the expected value associated with each option. Value is central to the subjective utility theory; in the medical context, the value associated with a decision can be assimilated to the therapeutic value of the decision to the patient. This may be defined in terms of improvement in the patient's quality of life, productivity, and fundamentals of the disease state.[3] Establishing such value is a complex process and may involve other variables such as the period of time over which any combination of the above is observed or expected, the social impact, further discussions with the patient, etc.

Typically, decision under uncertainty is common in the medical field; as a result, clinicians often rely on the expected health benefit to navigate the possibilities given the circumstances. For example, telling a patient that they are suffering from cancer when they are not may have serious consequences. Similarly, not being able to tell a patient that they are suffering from cancer when they are may be equally bad. However, irrespective of the circumstances and even when substantial evidence against the presence or absence of the disease is lacking, the medical expert is still expected to produce a diagnosis, and this with a certain level of confidence.

---

[3]Value in healthcare may include many other components, including the monetary cost associated with providing the service. But we believe that a patient-centric approach to value is more appropriate for our discussion.

In effect, the medical diagnosis process is a complex process requiring the gathering, collection, integration, and interpretation of a large variety of information to determine a potential diagnosis (National Academies of Sciences, Engineering, and Medicine 2015). Specifically, this information, all of which is critical for making a successful diagnosis of what the patient is suffering from, may include the patient's information, symptoms, and history. However, the care professional's familiarity with the patient's symptoms cannot be underestimated and is probably just as important. Stated differently, the medical professional's experience can significantly influence the diagnosis given to the patient. The importance of the medical professional's experience is well documented and can be generalized under the concept of evidence-based medicine as explained in the following (Sackett et al. 1996):

> [...] Evidence based medicine is not "cookbook" medicine. Because it requires a bottom-up approach that integrates the best external evidence with individual clinical experience and patient's choice, it cannot result in slavish, cookbook approaches to individual patient care. External clinical evidence can inform, but can never replace, individual clinical expertise, and it is this expertise that decides whether the external clinical evidence applies to the individual patient at all and, if so, how it should be integrated into a clinical decision. [...] By individual clinical expertise we mean the proficiency and judgement that individual clinicians acquire through clinical experience and clinical practice. Increased experience is reflected in many ways, but especially in more effective and efficient diagnosis [...].

As highlighted by Sackett, some of the knowledge required for effective and efficient decision-making in the medical field may come with years of practice, especially for rare diseases or diagnoses requiring the analysis of complex medical images such as magnetic resonance imaging (MRI) and computerized tomography (CT) scans (Nguyen et al. 2017; Salkowski and Russ 2018). Unfortunately, even for experienced medical professionals, such experience is somewhat limited to the practitioners' own experience and ability to access external evidence. This to some extent may weaken the medical professional's confidence in their decision-making. More precisely, it may increase the uncertainty in the diagnosis established by the medical professional. Such uncertainty is further amplified when a patient presents undifferentiated symptoms that change over time (Hatch 2016; Simpkin and Schwartzstein 2016). One of the many manifestations of such uncertainty is diagnostic variation, where medical professionals give different diagnosis to the same patient. However, it is important to highlight that physicians are generally aware of these limitations and routinely seek or recommend that patients seek additional opinions (wherever appropriate) on diagnosis they have given, especially during the early stages of certain diseases.

As discussed earlier in this book, because of its ability to process large volumes of data, AI can harness years of experiences through clinical research data to provide a diagnosis hint and therefore assist physicians in the diagnostic process. Stated differently, AI models are predictive, meaning that when they are provided with enough relevant data on a disease, they can classify a patient as diseased or not diseased, given the affected patient's input. Machine learning algorithms that perform this kind of tasks are known as classification algorithms. Such algorithms, as their name suggests, are typically used for classification problems discussed in the "Supervised Problems" section of Chapter 5. In simple terms, a classification algorithm attempts to classify its input into one of the available labels. An interesting property of some of these algorithms is that they can often, in addition to predicting the label, also predict the probability of the respective label. Such probabilities can be interpreted as the confidence level on the prediction depending on the algorithm. Precisely, some algorithms such as logistic regression[4] produce probabilities that can be directly interpreted as probabilities of the respective label. Other algorithms such as neural networks only produce a poor estimate of the probabilities. Such an estimate needs to be adjusted to reflect the model's confidence level on the prediction, that is, the probability of the predicted label. Importantly, irrespective of the classification algorithm used, one can obtain the probability of the respective label and therefore quantify the uncertainty associated with the prediction.

The above suggests that an AI algorithm, when carefully designed, can help quantify the uncertainty associated with the diagnostic decision-making provided that all alternative choices are properly defined up front. Without loss of generality, hypotheses on the presence or absence of the disease constitute a good starting point. In this setting, the medical expert is interested in determining whether the patient is diseased or not. To illustrate this process, let us consider the example of a classification algorithm that aims to classify a patient into two categories, diabetic and nondiabetic, given their age, body mass index (BMI), and blood pressure. For the sake of this discussion, let us assume that the patient is a 27-year-old female with a BMI of 24 and has a blood pressure of 97/65 mmHg. Using the information learned during its training, the algorithm determines the probability of the patient being diabetic to be 0.78.[5] Because probabilities must sum up to one, this means that the probability of the patient not being diabetic is 0.22. Given these probabilities, we can claim that the algorithm is 78% confident that the patient of whom the input was provided can be classified as diabetic.

---

[4]Logistic regression is an ML algorithm used to estimate the probability of an event occurring given some previous information.
[5]We assume that this probability can be interpreted as the confidence level of the algorithm.

Reflecting on the key concepts of the subjective utility theory (decision under uncertainty, value, and probability), the above discussion suggests that AI can effectively be used to inform the "probability" component and therefore support the decision-making in the medical context. More generally, when provided with enough information and appropriate context, AI can help estimate probabilities associated with one event or another and therefore facilitate and support decision-making in the medical context.

Beyond providing support and helping improve medical decision-making, AI presents an attractive proposition for the management and prevention of certain diseases. Particularly, chronic conditions can benefit from this as they require both accurate diagnosis and assessment and optimization of medication. In addition, AI along with associated technologies could help with the monitoring, diagnosis, detection, and identification of early signs or symptoms of certain diseases before the chronic phase of the disease is set. Such a preventive approach to healthcare is necessary not only to prevent people from getting ill in the first place but also to help patients better manage and minimize or slow down the disabling effects of chronic diseases. For example, monitoring and predictive analytics can help prevent complications associated with health conditions such as stroke or a fracture.

From a financial perspective, the predictive nature of AI is extremely attractive because the cost of treating a disease is generally significantly lower than the rehabilitation cost for the same disease. This is particularly true for chronic health conditions such as diabetes, cancer, and heart disease. For example, in 2017 in the United States alone, the estimated medical costs and loss of productivity of diagnosed diabetes were more than $300 billion. In this setting, AI can help estimate the likelihood of the patient developing the disease given the right context. Such information can then be relied upon to provide affected individuals with the help they need.

Medical Decision-Making and AI: The Use of Quantitative Prediction

As discussed in the previous section, when provided with the right information, AI can be used to estimate the probability of a patient respectively having a disease and not having the disease. These probabilities can be relied upon to determine whether a patient is diseased or not and often form the basis of AI diagnosis. This section discusses two studies on the prediction of medical decision-making, looking at their methodology and results.

Predicting Diabetic Retinopathy in India

A recent study conducted by a group of researchers focused on assessing automatic interpretation of retinal fundus photographs for large-scale screening and detection of diabetic retinopathy (DR) in India (Gulshan et al. 2019). DR was chosen because of its high prevalence in India. According to a 2009 report, approximately 18% of India's urban population affected by diabetes suffers from DR (Raman et al. 2009). With an annual incidence and progression ranging from 2.2% to 12.7% and 3.4% to 12.3%, respectively (Charumathi et al. 2019), DR is the leading cause of visual impairment worldwide. For the sake of clarity of the use case below, it is important to have a shared definition of DR. It is defined by the UK National Health Service (NHS)[6] as a complication of diabetes caused by high blood sugar levels damaging the back of the eye (retina). Like many chronic diseases, if DR is detected in an early stage, it can be treated effectively. Typically, screening for DR involves taking photographs of the retina once the pupils of the eye have been dilated with some medication. Photographs taken are then analyzed by an expert to determine the patient's risk level. The study presented in this section examined how the analysis of these photographs can be automated through AI.

The methodology of the study focused on using AI to analyze retinal fundus photographic images for the presence or absence of diabetic retinopathy. The study input data was retinal fundus photographic images obtained from patients of two tertiary eye care centers in South India. Over 3000 patients were included in the study. These patients were at least 40 years old and have previously been diagnosed with diabetes. Excluded from the study were patients with a history of intraocular surgery other than cataract surgery; ocular injections for DME or proliferative disease; a history of any other retinal vascular disease, glaucoma, or other diseases that may affect the appearance of the retina or optic disc; medical conditions that would be a contraindication to dilation; overt media opacity; or gestational diabetes (Gulshan et al. 2019). For each of these patients, the study performed a non-mydriatic fundus photography. Using the International Clinical Diabetic Retinopathy (ICDR) scale (Wilkinson et al. 2003), the study then had the captured fundus photography manually graded by trained graders and retinal specialists. Trained graders had between 7 months and 5 years of experience, whereas retinal specialists had between 15 months and 10 years of experience. The study then adjudicated all graded images by three senior retinal specialists and resolved disagreements accordingly. The final output target was a binary classification task aiming to determine whether an image was an instance of referable diabetic retinopathy (DR) or an instance of referable diabetic

---

[6]National Health Service (NHS) is the umbrella term for the publicly funded healthcare systems of the UK.

macular edema (DME). In this setting, an image was considered referable for DR if it had moderate or worse DR, whereas it was considered for DME if it had a hard exudate within 1 disc diameter of the macula. A deep neural network model[7] was respectively trained on more than 100,000 images, tuned on approximately 40,000 and validated on approximately 5500 fundus photography images to produce a system that grades fundus photography images according to the ICDR; however, only the above outcomes (DR and DME) were considered by the study (Gulshan et al. 2019).

Performance of the trained model expressed in terms of sensitivity and specificity were comparable to that of trained graders and retinal specialists. Precisely, the model's sensitivity across two sites was 88.9% and 92.1% for moderate DR, and 97.4% and 95.2%, respectively, for DME, whereas the specificity across the same sites was 92.2% and 95.2% for moderate DR, and 90.7% and 92.5%, respectively, for DME (Gulshan et al. 2019). A summary of the overall performance evaluation conducted by the authors of the study can be seen in Figure 8-1, where CI stands for confidence interval and proposes a range of plausible values for each of the performance metrics, namely, sensitivity and specificity. Despite the relatively small validation data set used in this study, results achieved by the authors are promising and show that AI can be effectively used to assist clinicians in the task of screening for early signs and detection of diabetic retinopathy. This study is not the only attempt to demonstrate the effectiveness of AI for the early detection of DR, a similar study was conducted by Google Health in Thailand (Beede et al. 2020).

---

[7]Deep neural networks are a class of ML algorithms that attempts to mimic the biological structure and functioning of the brain for solving complex tasks.

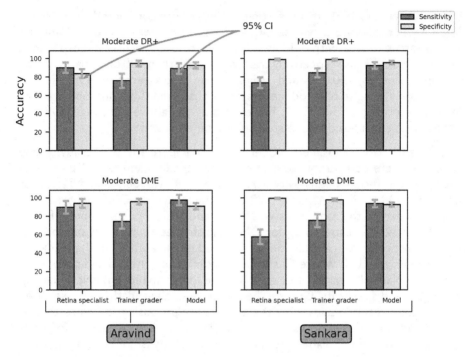

Figure 8-1. Overall summary of the performance of the system developed in the frame of the diabetic retinopathy study (Gulshan et al. 2019)

Predicting Cardiac Arrest

Another group of researchers focused on the diagnosis of heart failure (HF) using AI (Choi et al. 2020). The goal of the study was to evaluate the accuracy of the diagnosis of HF by an AI-powered clinical decision support system (CDSS). The rationale behind the study was that if the effectiveness of a system such as the one assessed by the study can be proven, then they could assist in the diagnosis of HF, especially in the absence of a HF specialist. This is unfortunately often the case, especially in developing countries where the health workforce per 10,000 population is extremely low (World Health Organization 2021). Affecting more than 26 million people worldwide, HF has been declared a global pandemic (Ponikowski et al. 2014). In simple terms, HF occurs when the heart is unable to pump blood around the body properly often because the heart is becoming too weak or stiff. There are typically three types of HF depending on the amount of blood the left ventricle pumps out with each concentration, also referred to as "ejection fraction." These are

- Heart failure with reduced ejection fraction (HFrEF) or systolic failure

- Heart failure with preserved ejection fraction (HFpEF) or diastolic failure

- Heart failure with mid-range ejection fraction (HFmEF)

The study assessed the performance of the AI-based CDSS system on the above-listed types of HF against that of HF specialist and non-heart failure specialist.

To achieve this aim, the study input included a retrospective cohort of 1198 patients with and without heart failure. Data from approximately half of these patients were used to train the system and the other half for testing it. Using the training set, the authors of the study created an AI-based CDSS system that combines an expert-driven CDSS system and an ML-driven CDSS system. The expert-driven system is made up of rules defined by HF experts for the detection of various types of HF, whereas the ML-driven system is composed of rules generated using an ML algorithm referred to as the classification and regression tree (CART).[8] The main features selected by the ML algorithm included the left ventricle ejection fraction, atrial volume index, and left ventricle mass index. The output target was a multiclass classification task as to whether the patient had HFrEF, HFpEF, HFmEF, or no HF. Following the training, the model was tested on the remaining 598 patients that were not included in the training. Out of these patients, 490 had HF and 108 did not.

The model obtained a sensitivity and specificity of 0.94 and 0.99, respectively. The study also compared results of the AI-based CDSS system with those of experts and found that they were comparable. Precisely, the study showed that the concordance rate between the AI-based CDSS system and the HF specialist was 100% for patients with HFrEF and HFmEF, 99.6% for patients with HFpEF, and finally 91.7% for patients with no heart failure (Choi et al. 2020).

Despite the small sample size that was used for evaluating the performance of the AI system presented in this study, it shows that AI can be used for detecting or identifying early signs of heart failure. Additionally, this study also illustrates how human knowledge can be integrated into the design of AI systems not only to improve their accuracy but, more importantly, for greater transparency.

Overall, the above studies, despite their apparent limitations, demonstrate the undeniable potential of AI in the healthcare sector. However, many challenges lie ahead of the road toward the widespread adoption and deployment of those AI systems in real-life settings. Some of these challenges are discussed in the next section.

---

[8]CART is a term commonly used to refer to decision tree algorithms that can be used for classification or regression tasks.

Limitations of AI in Decision-Making in the Medical Context

Thus far in this chapter, we have discussed AI's ability to support decision-making in the medical context. Precisely, we showed that the predictive nature of AI can play a fundamental role in assisting decision-making in the medical context. This is because AI can efficiently calculate the likelihood of any given predefined outcome provided the right circumstances. It achieves this by relying on the knowledge extracted from available historical information related to the problem at hand. So, AI can virtually assist any medical decision given the right circumstances. Yet, only limited progress has been observed in the adoption of AI systems in real-life medical settings.

This section presents some of the challenges or limitations that hinder the large-scale deployment or adoption of AI in the medical context.

Legal Responsibilities

Legal responsibility is an ongoing debate around the adoption and use of AI-assisted systems in critical settings such as clinical decision-making. Although legal responsibility involves many other components, the biggest challenge is probably related to the designation of liabilities. From a holistic perspective, an AI-assisted decision-making system is not limited to the AI system itself but involves three key components:

- The AI system, that is, the software performing AI and related functions

- The designer or developer, that is, the individual or organization that developed the system

- The users, that is, medical professionals/clinicians who will be using the system

An effective responsibility model for AI must consider all these three components where appropriate when allocating responsibilities or, more precisely, liabilities. In essence, it can be argued that the software component cannot be responsible for anything it does because its behavior is fully predetermined by its programming. This programming could be either of the following types:

- Explicit, that is, the designer explicitly defines the sequence of instructions that the software will perform to solve the problem

- Implicit, meaning that the designer simply provides some context, and the software figures out how to solve the problem

Irrespective of the programming model, the algorithm's behavior is entirely prescribed by the programmer. To better understand this, let us take a step back and revisit how ML models are programmed. Typically, the designer provides data, a learning algorithm, and any other optimization parameter; the learning algorithm then uses the information it has been given to create a model or learned algorithm, which is a solution to the problem at hand. It can be argued that the programmer may not have been the one who wrote the learning algorithm. However, the responsibility remains in the hands of the designer because they have a free choice to choose any other learning algorithm. This points to the importance of the designer having a reasonable understanding of the algorithm as well as its limitations.

Now that we have established that the software cannot bear any responsibility, it leaves the designer and the users of the system. As alluded to, the algorithm's behavior is prescribed by its programming. Consequently, it is logical that the designer or the organization that created the algorithm is liable for any poor behavior that the software may exhibit. Stated differently, by developing and selling AI systems that assist decision-making in the medical context, an organization implicitly becomes a member of the medical profession and therefore acquires the responsibility to understand the complexity of decision-making in this context and has the duty to operate according to values and principles of the medical profession.

Although the liability related to the behavior of the AI system is typically allocated to the designer, the clinician has the duty to seek clear understanding of the software and the implications of its performance in practice. Precisely, ML models are not perfect or 100% accurate, meaning that there is always an element of uncertainty involved in any prediction they make. Consequently, in these circumstances, the clinician must use the predictive system with reasonable care. In this setting, what constitutes a reasonable care should be defined by the relevant medical authority. For example, the Standard of Care[9] could serve as the reference for this.

Lack of Standards and Trust

The limited or lack of trust in AI-assisted decision-making in the medical context is arguably one of the biggest challenges that AI faces today. This lack of trust can be attributed to two main factors:

- The complex nature of AI, which makes it difficult to understand the inner working of AI-assisted decision-making systems

---

[9]The Standard of Care can be defined as informal or formal guidelines that are generally accepted in the medical community for the treatment of a disease or condition (Oberman 2017).

- The lack of effective standard for the development, evaluation, and deployment of AI systems

The complexity of how certain classes of AI algorithms arrive at their decisions constitutes a barrier for their adoption in critical environments such as the clinical environment where a decision can directly impact a patient. Such class of algorithms, including deep neural networks, more than often is ironically the one that leads to the most accurate and promising results[10] in studies around AI-assisted decision-making in the clinical context. This means that in practice it would be impossible for a clinician to tell with certainty how the decision made by the algorithm was achieved. This is problematic for any system that is intended to support and/or improve decision-making. Assuming the clinician fully trusted the decision suggested by the AI system, they would probably struggle to explain that decision to the patient who is expected to be involved in treatment decisions. A much more complex scenario may arise when the system's decision diverges from the conclusion established by the clinician. Extensive effort is being devoted by the AI community to make AI decision logic more understandable.

Another problem with AI in particular and digital technology in general is the lack of standards for the development and validation of AI systems for deployment in the medical environment. Such standards would, for example, prescribe strict guidelines and methodologies for the evaluation of AI-assisted technologies for use in the medical context. Importantly, a third party or regulator should be able to verify that these standards have been followed by any organization developing AI-assisted decision-making systems for the clinical environment.

When it comes to healthcare, people are generally more skeptical to innovation and new things. Specifically, patients are unlikely to trust AI-assisted decision-making systems when their design, development, and evaluation are lacking appropriate regulatory oversight. For example, at the time of this writing, the coronavirus disease 2019[11] (COVID-19) pandemic has caused more than two million deaths worldwide and destabilized many economies. Despite everything spinning out of control, the COVID-19 vaccine has been received with mistrust worldwide in part due to its record development time. In effect, the public is concerned that the safety of the vaccine was compromised to speed up the development and testing processes. In response, various government agencies are undertaking massive campaigns to reassure the public that the COVID-19 vaccine is safe and has been extensively tested (UK Department of Health and Social Care 2021). This engagement from public

---

[10]Generally, the more complex an AI algorithm is, the more accurate it is, because it can efficiently identify hidden and complex patterns.

[11]Coronavirus disease 2019 (COVID-19) is a contagious disease caused by severe acute respiratory syndrome coronavirus 2 (SARS-CoV-2).

authorities can be illustrated by Figure 8-2 which shows a COVID-19 vaccine advertisement poster found in a street of London, UK.

Figure 8-2. Poster from a street of London encouraging the public to take the COVID-19 vaccine

COVID-19 vaccine concerns remain despite regulatory agencies such as the Food and Drug Administration (FDA) in the United States and the Medicines and Healthcare products Regulatory Agency (MHRA) in the UK approving the vaccine for public use. It is important to highlight that such regulatory agencies are there to ensure that drugs and vaccines are rigorously tested and their effectiveness appropriately assessed. Yet, at the time of this writing, the public remains reluctant.

While the COVID-19 vaccine is different from AI-based decision-making systems, some parallels can be drawn between them. They are both innovative, lack public understanding, and in need of public trust. But, more importantly, it provides a glimpse into the resistance that practical adoption of AI is likely to face not only in the clinical context but in any environment where the public believes safety should be a priority. A resistance that is likely to be fortified by the absence of an effective regulatory arrangement that[12] safeguards public interest and safety. Additionally, the use of AI in the medical environment does not have the backing of public institutions and authorities as does the COVID-19 vaccine, suggesting that the road to widespread use or adoption of AI in the medical decision context is expected to be much tougher. However, lessons can be learned and appropriate actions taken to avoid being in a situation similar to that raised by the COVID-19 vaccine in the future.

A common limitation of AI studies in the medical decision setting is related to the discussion around reproducibility. Reproducibility can be defined as the ability to reproduce the results achieved by an observational or experimental study. This represents a big challenge because most published studies examining the use of AI in the clinical context are not reproducible and therefore can be argued to lack proper validation. Not being able to reproduce AI studies is problematic for the design of systems for use in advance patient care and does not build confidence among clinicians and scientists.

Multiple factors contribute to poor reproducibility of AI studies in the medical decision context. However, the most important factor is arguably the fact that researchers generally do not make the source code and data used for their studies available to the public. This can be attributed to multiple reasons, such as the need to maintain an advantage over competitors and the nature or origin of the data used by the study. In effect, in the medical decision context, the data introduces an extra level of complexity that is discussed below.

Medical Data

AI studies in the medical decision context are generally limited by the lack of data to reproduce related experiments. This can be attributed to the sensitive nature of medical data, which in the wrong hands could be misused if not harm affected patients. It is therefore not surprising that little to no medical-related data is shared among organizations. Although sharing such data could

---

[12]An effective regulatory arrangement should be able to efficiently perform all three components of regulation discussed in Chapter 7.

help advance AI research and demonstrate the feasibility of past studies, organizations must do everything in their power to ensure that patients' data is kept private. Anonymization is often used in the attempt to maintain a patient's privacy. However, this does not always guarantee privacy, because de-identified data can be merged with other data sources to recover the identity of the patient (Savage 2016).

Because of issues such as bias introduced and discussed in Chapter 2 and the "Beyond Traditional AI Performance Metrics" section of Chapter 5, current medical data may also be inappropriate for the development of AI tools for assisting decision-making. For example, a study conducted in the United States showed racial bias in pain assessment and treatment recommendation (Hoffman et al. 2016). Similar issues may be observed in other areas of healthcare. This means that any AI system built on top of this data will exhibit the same flaws. Consequently, such data is inadequate for the development of AI systems for assisting decision-making in the medical context. Additionally, AI models often do not generalize/perform well beyond the population from which the training data was collected. This is problematic because it may underrepresent some of the patients. Issues related to the completeness of the data may also affect the performance of the AI system in the medical context. For example, doctors often come to a diagnosis after multiple visits from a patient and other information that may be poorly documented or captured by the information management system.

Another data-related issue often observed in AI studies in the medical decision context revolves around the fact that the data used to train the AI system presented often does not reflect the kind of data a clinician would observe on a typical day. Specifically, the data used in studies often do not present the challenges that are observed in real-world clinical environments. Consequently, the advertised performance of models discussed in such studies is generally misleading and does not tell the whole story. For example, Google Health's retinopathy AI system laboratory testing indicated it could identify early signs of diabetic retinopathy with more than 90% accuracy. However, during testing in real-world settings in Thailand, the system did not live up to expectations and was to some extent impractical in part because it was designed to reject fundus images falling below a certain threshold of quality. Put simply, retinal images used for training the system were selected to maximize the system's classification accuracy (Beede et al. 2020).

The above example highlights the importance of defining performance metrics in terms of business objectives (see Chapter 5 for more details), which in the medical context would be improving outcome and/or satisfaction for the patients. Consequently, this outcome may involve several components, some of which are contextual and irrelevant to the AI system's accuracy.

Although the real-world deployment of Google Health AI retinopathy solution was not a complete success, there is great value in publishing studies like the one conducted by Google Health. The example illustrates to which extent studies examining the use of AI systems in the medical decision context are disjointed from the real world. Precisely, assumptions made when designing and testing an AI system in the laboratory did not hold in practice. This may lead, as was the case for the Google Health real-world testing, to poor outcomes for patients and overall frustrating experience both for patients and clinicians. More importantly, this example provides a glimpse into the complexity involved in deploying AI systems in real-world clinical environments.

Conclusions and Discussion

Healthcare professionals and patients must make decisions under uncertainty daily. In this chapter, we discussed the relevance of AI-assisted decision-making in the medical or clinical context. The discussion on decision-making provides some clarity on why AI within the medical decision context should primarily be perceived as a tool for helping clinicians structure their decisions. Precisely, we show that artificial intelligence can be used to quantify the uncertainty in the medical decision context by estimating the probability of an outcome given the circumstances.

Through two studies, we illustrated advances that have been made thus far and discussed potential challenges that medical decision-making assisted by AI may face in real-life clinical environments. Such challenges revolve around the responsibility model associated with AI systems in particular and digital technologies in general, the lack of standards and trust in AI systems, and finally the complex nature and challenges related to medical data.

It is common for studies examining the use of AI in medical decision context to claim that the proposed AI model either outperforms or performs at a level comparable to that of expert clinicians. Despite steady progress being made in AI in relation to clinical decision-making, such claims are generally misleading, because the corresponding studies lack any kind of evaluation/ testing in real-world clinical environments. As discussed in the "AI Performance Metrics Overview" section of Chapter 5, AI systems' performance is generally assessed using traditional performance metrics;[13] this is typically the case for AI systems designed to operate in clinical settings. However, these metrics often do not reflect all objectives. This means, for example, that there is no guarantee that the corresponding AI system can or will improve the outcomes for the patient once deployed in a real-world environment. In effect, the only requirement for the AI system is to perform well with respect to traditional performance metrics. This can be illustrated by the Google Health example, which despite the 90% accuracy fell short in real-world testing. However, the

---

[13]We use the term traditional performance metrics to refer to AI performance evaluation metrics such as accuracy, sensitivity, specificity.

Google Health team should be commended for reminding the AI community through this study that establishing performance metrics such as accuracy in lab settings does not mean much without real-world clinical testing.

Additionally, we learned in "The Need for AI in Medical Decision-Making" section that the concept of "value" is fundamental to decision-making in the real-world medical decision context. Understood as the benefit that the decision will bring to the patient, value cannot be fully modeled through traditional performance metrics and may vary depending on the local context. In summary, evaluation in real-world clinical settings must be an integral part of the design and development of AI systems for the medical decision context.

Conclusions and Discussion

Over the past few years, there has been an increasingly enthusiastic interest in artificial intelligence (AI) or data science (DS). This is because DS impacts almost every aspect of our lives and often outperforms traditional approaches in solving complex problems we face daily. Consequently, an increasing number of organizations all around the globe are devoting significant efforts to incorporate AI in their operational strategy as they continue to witness its transformative potential. However, realizing this potential often gives rise to unforeseen ethical consequences including, but not limited to, fairness/bias, trust, transparency, and privacy.

Building a sustainable culture for developing and deploying AI systems starts with identifying and mitigating the abovementioned consequences. The sustainable AI framework (SAIF) achieves those goals through its process – controls – governance operating model and helps organizations

- Size the impact of artificial intelligence on their ability to create value in the short, medium, and long term

G. L. Tsafack Chetsa, *Towards Sustainable Artificial Intelligence*,
https://doi.org/10.1007/978-1-4842-7214-5_9

- Understand the elements that enable artificial intelligence along with their complexity and prepare for future regulations

- Audit their AI systems to assess their current risk profile and exposure

SAIF accomplishes the above by integrating the social, economic, and political implications of AI systems as inherent aspects of their design and development. This involves a clear understanding of the implications of the fundamental elements of AI systems. Typically, such elements are the data and the associated inferencing capability. On the one hand, data constitutes a business asset as organizations can rely upon to create innovative products. On the other hand, it easily becomes a liability if standards related to its privacy, security, and compliance with regulations are not met. Like the data element of AI systems, AI inferencing capability or more generally the predictive nature can be detrimental not only to the organization providing the associated service but more importantly to those using such a service.

The operating model of the SAIF framework, which is referred to as the process – controls – governance model, provides a tool for controlling the behavior of AI systems and therefore safeguarding against unwanted consequences or undesired outcomes. The process – controls – governance model is centered around the need to understand and assess an organization's risk profile and exposure while gaining a better understanding of the AI system either being used or under development. In its simplest form, the process – controls – governance approach consists of defining and implementing controls around the data science development process. The governance component of the model ensures that resources and structures are in place to oversee the development process and associated controls by establishing and maintaining accountability as well as delegating core oversight to individuals with the right skill sets.

AI systems are traditionally assessed only on their ability to infer an adequate outcome given new data. However, as AI continues to infiltrate various aspects of daily life, it is essential to assess their effectiveness against known concerns such as bias and privacy. These concerns are covered under the concept of soft performance metrics which should be monitored throughout the system's life cycle. Because soft performance metrics are often unrelated to the algorithm's ability to provide the desired outcome, involving a varied audience in the design and development of AI systems should help identify potential soft performance metrics and more success criteria for the AI system. Commonly, the lack of diversity within the abovementioned audience, coupled with poor communication between the business and the development team, often hinders the identification of relevant soft performance metrics.

While frameworks like SAIF enable organizations to develop AI systems that meet certain social standards, regulations extend the need to meet those standards into legislations to guarantee users' protection and safeguard the public against abusive behaviors. The effectiveness of the current regulatory arrangement is challenged by the lack of a proper mechanism for their adjudication. We argue that an effective regulatory arrangement in the context of AI must combine features of industry self-regulation and regulation by the government. Precisely, this can be achieved through a third-party regulatory market designed by both the government and the AI industry. Under such circumstances, the government defines appropriate legislation and oversees their enforcement, whereas the third-party market develops mechanisms and tools for the adjudication of the regulation. Pressure from the public or more precisely consumers of AI services will ultimately shape the government's attitude toward AI and consequently its view on the establishment of an effective regulatory arrangement for AI. However, it is in most organizations' best interest to self-regulate where regulations are lacking. In effect, AI continues to be commoditized (access, knowledge, and understanding), giving rise to alternative, often better, services.

The potential to help organizations generate revenues in some way or form is arguably one of the biggest, if not the biggest, motivations behind the race for innovation through AI (knowledge base or data and inferencing capability). This is illustrated by the continuous revenue stream that the data represents. This data for the most part results from the interaction between an organization and its customers. Currently, in exchange for this data, customers are rewarded by either marginal improvements on services they receive or are provided with the service free of charge. However, many would argue that the current revenue distribution scheme is skewed toward the organization collecting the data. Stated differently, customers should be properly compensated for the wealth they are helping organizations accumulate. Effective mechanisms for a fairer redistribution of revenues generated from data monetization may include establishing a tax on data. Another alternative may consist of offering customers the ability to sell data resulting from their interaction with one organization to competing organizations. Similar services already exist through data brokers, yet ironically customers create the data that is sold to a buyer who then uses it to create more opportunities for itself at the expense of the same customers. Stated like this, claiming that customers are being fooled in daylight is not exaggerated.

From assisting decision-making to helping streamline day-to-day operations, AI arguably has a lot to offer in the healthcare industry. However, the complexity of the data manipulated, coupled with the fact that most organizations attempting to provide AI-assisted services in this context are profit driven, calls for strict legislations and guidelines for the development and evaluation of AI applications in the industry. Failing to provide such guidelines may hinder the development of AI in this sector and place patients at greater risk and exposure to abusive practices.

Need for Standards and Definitions

Unlike traditional AI performance metrics, soft performance metrics are often subject to multiple interpretations. Bias, for example, occurs when the outcome of an algorithm is unfair to a specific group of individuals (see the "Soft Performance Metrics" section of Chapter 5). Avoiding bias may be a suitable business objective, and thus soft performance metrics are needed in order to achieve this objective. As our definition of bias points out, bias revolves around the concepts of "fairness" and "groups." This means that alternative approaches for defining these concepts may lead to different and potentially conflicting outcomes. In other words, while procedural approaches, including processes and technological tools, may help one to identify and mitigate bias, addressing bias requires a clear understanding of its components first. The multiplicity in the interpretation of fairness, for example, is well summarized by Dr. Orville Boyd Jenkins in his blog:

> Fairness can be interpreted as being equal in provision, in opportunity or in result. From each point of view, the other point of view may seem unfair.[1]

In following this logic, an AI system may be fair by some metrics and yet unfair by others.

Much of the conversation around bias has focused on fairness, when understood as group equality – that is, making the outcome of the AI model equitable across all groups. As a result, a system may be arbitrarily fair or unfair depending on the level of granularity at which one defines groups and group membership. Moreover, groups' definition or understanding may be influenced by cultural and geographic factors.

Another hurdle likely to arise in discussion around fairness is related to the fact that a "fair" statistical outcome does not guarantee that individuals will not be discriminated against. This can be attributed to the fact that statistics are defined with respect to group averages. Meaning that there is no guarantee that an individual who is not an "average" member of the group would be treated fairly. Attempts to remedy this consist of defining fairness at the individual's level. One approach suggests that fairness should be understood

---

[1] http://www.orvillejenkins.com/faithlife/fairnessfl.html

as "similar individuals are treated similarly" (Dwork et al. 2012). Stated differently, individuals who are similar should, approximately, receive equal outcomes. However, the biggest challenge lies in identifying or defining an appropriate similarity metric for the task at hand. Such similarity metric can be thought of as the extent to which two individuals are similar with respect to the task at hand.

Many organizations are trying to maintain an internal (yet contextual) definition/understanding of these concepts. However, policy makers or governments, along with civil society, are required to play a leading role in the establishment of a shared understanding of what fairness means. From what precedes, it is unlikely to establish a single and universal definition of fairness and how to assess fairness. As a result, different metrics and standards can be defined depending on the use case and the application domain. For instance, policy makers may prescribe standards or guidelines on how to define and/or assess an AI system for fairness in the healthcare industry.

This need for a common understanding and/or standards around fairness is probably best illustrated through the debate on data privacy and its evolution over time. For example, influenced by new technological advances, the European Data Protection Directive was created in 1995 introducing terms, including, but not limited to, processing, sensitive personal data, and consent. The Data Protection Directive was later replaced by the General Data Protection Regulation of 2018. Today, the debate has evolved into that of responsible management of personal data through mature IT governance, transparent processes, and modern applications. It will probably continue to evolve to take into account the influence of innovative technologies. The path from fairness to prescriptive metrics may take as much time as the data privacy journey; however, governments and civil society can help prioritize the multiple definitions of fairness.

While the above discussion is primarily focused on bias, the need for standards and shared understanding can be extended to soft performance metrics including interpretability and privacy as well as other aspects of AI. As illustrated by efforts being carried out by various working groups within the Institute of Electrical and Electronics Engineers (IEEE), the industry is well aware of the need for guidelines in the form of international standards.

The debate on privacy at its core is contextual. Consequently, such debate is influenced by views on what constitutes a person's rights and how it should be regulated. This is evidenced by current data protection and privacy regulations which are fragmented with diverging regional and national approaches. Additionally, in developing countries, they are nonexistent, and usually difficult to enforce when available, because of their social, economic, and political climate. Nonetheless, organizations wanting to engage in cross-border trades of data-related services must adapt to the data protection and privacy regulations of the environment in which they want to operate or sell their services. This creates an extra level of complexity for most organizations and may give rise to data privacy violation and other ethical issues discussed in this book. Worse, in developing countries, as highlighted, rules may not exist at all, meaning that adopting a responsible conduct on matters related to data protection and privacy is entirely left at the discretion of the organization delivering the service. A situation that most people would qualify as problematic. In summary, many issues may arise from divergences in data protection and privacy regulations. Such issues can be mitigated through an agreed set of data protection rules and principles.

Beyond the need for standards around the definition and specification of key concepts such as interpretability and fairness/bias discussed above, there is an increasing necessity to introduce an effective regulatory framework for the development and deployment of AI systems. Such a framework would enable the development of consistent and trustable AI systems. This is of particular importance in domains like the healthcare industry where questions of accountability and transparency are of great importance. More generally, in critical industries such as the healthcare industry, a consistent and legally binding distribution and definition of roles and responsibilities among AI systems components (software, designer, and users – see the "Legal Responsibilitiess" section of Chapter 8) is needed to fully realize the potential of AI while safeguarding the public. Importantly, prescriptive guidelines for the development and deployment of AI systems in critical industries will ensure that organizations have a clear understanding of the responsibilities that they are assuming by developing and/or selling AI systems that will operate in such industries.

Bibliography

Aditya, Krishna Menon, and C. Williamson Robert. 2017. "The cost of fairness in classification." *CoRR*.

Aldén, Lina, and Mats Hammarstedt. 2016. "Discrimination in the Credit Market? Access to Financial Capital among Self-employed Immigrants." *Kyklos* 69: 3–31.

Alizadeh, Elaheh, Samanthe Merrick Lyons, Jordan Marie Castle, and Ashok Prasad. 2016. "Measuring systematic changes in invasive cancer cell shape using Zernike moments." *Integrative Biology* 8 (11): 1183–1193. https://doi.org/10.1039/c6ib00100a.

Al-Rubaie, M., and J. M. Chang. 2019. "Privacy-Preserving Machine Learning: Threats and Solutions." *IEEE Security Privacy* 17 (2): 49–58.

Andrew, Brien. 1998. "Professional Ethics and The Culture of Trust." *Journal of Business Ethics* 17 (4): 391–409.

Avishek, Joey Bose, and William Hamilton. 2019. "Compositional Fairness Constraints for Graph Embeddings." *CoRR*.

Baldi, Pierre. 2011. "Autoencoders, Unsupervised Learning and Deep Architectures." In *Proceedings of the 2011 International Conference on Unsupervised and Transfer Learning Workshop – Volume 27*, 37–50. Washington, USA: JMLR.org.

G. L. Tsafack Chetsa, *Towards Sustainable Artificial Intelligence*,
https://doi.org/10.1007/978-1-4842-7214-5

Bartlett, Robert, Adair Morse, Richard Stanton, and Nancy Wallace. 2019. *Consumer-Lending Discrimination in the FinTech Era.* Working Paper, National Bureau of Economic Research, National Bureau of Economic Research. Accessed May 02, 2012. doi:10.3386/w25943.

Beede, Emma and Baylor, Elizabeth, Fred Hersch, Anna Iurchenko, Lauren Wilcox, Paisan Ruamviboonsuk, and Laura M. Vardoulakis. 2020. "A Human-Centered Evaluation of a Deep Learning System Deployed in Clinics for the Detection of Diabetic Retinopathy." *2020 CHI Conference on Human Factors in Computing Systems.* Honolulu: Association for Computing Machinery. 1–12.

Beede, Emma, Elizabeth Baylor, Fred Hersch, Anna Iurchenko, Lauren Wilcox, Paisan Ruamviboonsuk, and Laura M. Vardoulakis. 2020. "A Human-Centered Evaluation of a Deep Learning System Deployed in Clinics for the Detection of Diabetic Retinopathy." *CHI Conference on Human Factors in Computing Systems.* Honolulu, HI, USA: Association for Computing Machinery. 1–12. Accessed March 31, 2021. www.technologyreview.com/2020/04/27/1000658/google-medical-ai-accurate-lab-real-life-clinic-covid-diabetes-retina-disease/.

Beutel, Alex, Chen Jilin, Zhao Zhe, and Ed H. Chi. 2017. "Data Decisions and Theoretical Implications when Adversarially Learning." *CoRR.*

Blagus, Rok, and Lara Lusa. 2013. "SMOTE for high-dimensional class-imbalanced data." *BMC Bioinformatics.*

Bo, Jiang, Zhang Ziyan, Lin Doudou, and Tang Jin. 2018. "Graph Learning-Convolutional Networks." *CoRR.*

Bonawitz, Keith, Hubert Eichner, Wolfgang, Huba, Dzmitry, Ingerman, Alex, Ivanov, Vladimir Grieskamp, Chloe Kiddon, Jakub Konecny, Stefano Mazzocchi, H. Brendan, Van Overveldt, Timon McMahan, David Petrou, and Ramage. 2019. "Towards Federated Learning at Scale: System Design." *SysML 2019.*

Brooke, Auxier, Rainie Lee, Anderson Monica, Perrin Andrew, Kumar Madhu, and Erica Turner. 2019. *Americans and Privacy: Concerned, Confused and Feeling Lack of Control Over Their Personal Information.* Pew Research Center.

Campitelli, Guillermo, and Fernand Gobet. 2010. "Herbert Simon's decision-making approach: investigation of cognitive processes in experts." *Review of General Psychology* 354–364.

National Academies of Sciences, Engineering, and Medicine. 2015. "The Diagnostic Process." Chap. 2 in *Improving Diagnosis in Health Care.* Washington (DC): National Academies Press (US).

Carroll, Noel. 2014. "In Search We Trust: Exploring how Search Engines are Shaping Society." *International Journal of Knowledge and Learning* 5: 12–27.

Char, Danton S., Nigam H. Shah, and David Magnus. 2018. "Implementing Machine Learning in Health Care – Addressing Ethical Challenges." *The New England Journal of Medicine* 378 (11): 981–983. https://pubmed.ncbi.nlm.nih.gov/29539284.

Charumathi, Sabanayagam, Banu Riswana, Li Chee Miao, Lee Ryan, Xing Wang Ya, Tan Gavin, B. Jonas Jost, et al. 2019. "Incidence and progression of diabetic retinopathy: a systematic review." *The Lancet Diabetes & Endocrinology* 7 (2): 140–149.

Chawla, Nitesh V., Kevin W. Bowyer, Lawrence O. Hall, and W. Philip Kegelmeyer. 2002. "SMOTE: Synthetic Minority over-Sampling Technique." *Journal of Artificial Intelligence Research* 321–357.

Choi, Dong-Ju, Jin Joo Park, Taqdir Ali, and Sungyoung Lee. 2020. "Artificial intelligence for the diagnosis of heart failure." *npj Digital Medicine.*

Oxford Learner's Dictionaries. n.d. *Oxford Learner's Dictionaries.* Accessed November 28, 2020. www.oxfordlearnersdictionaries.com/definition/english/regulation_1.

Dwork, Cynthia, Moritz Hardt, Toniann Pitassi, Omer Reingold, and Richard Zemel. 2012. "Fairness through Awareness." *Innovations in Theoretical Computer Science.* Cambridge, Massachusetts: Association for Computing Machinery. 214–226.

Dzindolet, Mary T., Scott A. Peterson, Regina A. Pomranky, Linda G. Pierce, and Hall P. Beck. 2003. "The role of trust in automation reliance." *Int. J. Hum.-Comput. Stud.* 697–718.

Feldman, Michael, Sorelle Friedler, John Moeller, Carlos Scheidegger, and Suresh Venkatasubramanian. 2014. "Certifying and removing disparate impact."

Feldman, Ronen, and James Sanger. 2006. *Text Mining Handbook: Advanced Approaches in Analyzing Unstructured Data.* New York, NY, USA: Cambridge University Press.

Finale, Doshi-Velez, and Kim Been. 2017. "Towards A Rigorous Science of Interpretable Machine Learning."

Ford, Liz. 2007. "So you want to work in ..." *The Guardian.* December 1. Accessed December 30, 2020. www.theguardian.com/money/2007/dec/01/workandcareers.graduates1.

Ganesan, N., K. Venkatesh, M. A. Rama, and A. Malathi Palani. 2010. "Application of Neural Networks in Diagnosing Cancer Disease using Demographic Data." *International Journal of Computer Applications* 1: 76–85.

Gillespie, Tarleton. 2017. "Algorithmically Recognizable: Santorum's Google problem, and Google's Santorum problem." *Information, Communication and Society* 20 (1): 63–80.

Goldman, E. 2008. "Search Engine Bias and the Demise of Search Engine Utopianism." In *Web Search: Multidisciplinary Perspectives*, 121–133. Springer Berlin Heidelberg.

Gulshan, Varun and Rajan, Renu and Widner, Kasumi and Wu, Derek and Wubbels, Peter and Rhodes, Tyler and Whitehouse, Kira, Marc Coram, Greg Corrado, Kim Ramasamy, Rajiv Raman, Lily Peng, and Dale Webster. 2019. "Performance of a Deep-Learning Algorithm vs Manual Grading for Detecting Diabetic Retinopathy in India." *JAMA Ophthalmology*.

Gupta, Anil K., and Lawrence J. Lad. 1983. "Industry Self-Regulation: An Economic, Organizational, and Political Analysis." *The Academy of Management Review* (Academy of Management) 8 (3): 416–425.

Hamish, Fraser, Coiera Enrico, and Wong David. 2018. "Safety of patient-facing digital symptom checkers." *The Lancet* 392 (10161): 2263–2264.

Hatch, Steven. 2016. *Snowball in a blizzard: a physician's notes on uncertainty in medicine.* Basic Books.

He, Haibo, and Yunqian Ma. 2013. *Imbalanced Learning: Foundations, Algorithms, and Applications.* Wiley-IEEE Press.

Hendricks, Vincent Fella, Vestergaard Mads, and Marker Silas. 2018. "Digital colonialism on the African continent." *IOL Business Report.*

Hipp, Jochen, Ulrich Guntzer, and Gholamreza Nakhaeizadeh. 2000. "Algorithms for Association Rule Mining – A General Survey and Comparison." *SIGKDD Explor. Newsl* (ACM) 2 (1): 58–64.

Hoffman, Kelly M., Sophie Trawalter, Jordan R. Axt, and M. Norman Oliver. 2016. "Racial bias in pain assessment and treatment recommendations, and false beliefs about biological differences." *Proceedings of the National Academy of Sciences of the United States of America* 4296–4301.

Hornik, Kurt, Bettina Grün, and Michael Hahsler. 2005. "arules – A Computational Environment for Mining Association Rules and Frequent Item Sets." *Journal of Statistical Software* 14.

Hosmer, Larue Tone. 1995. "Trust: The connecting link between organizational theory and philosophical ethics." *The Academy of Management Review (AMR)* 379–403.

Hu Zhang, Brian, Lemoine Blake, and Margaret Mitchell. 2018. "Mitigating Unwanted Biases with Adversarial Learning." *CoRR.*

Jordan, Frith. 2017. "Invisibility through the interface: the social consequences of spatial search." *Media, Culture & Society* 39 (4): 536–551.

Kamishima, T., S. Akaho, and J. Sakuma. 2011. "Fairness-aware Learning through Regularization Approach." *2011 IEEE 11th International Conference on Data Mining Workshops.* 643–650.

Kosinski, Michal, David Stillwell, and Thore Graepel. 2013. "Private traits and attributes are predictable from digital records of human behavior." *Proceedings of the National Academy of Sciences* (National Academy of Sciences) 110 (15): 5802–5805.

Krawczel, Peter, and Rick Grant. 2009. "Effects of cow comfort on milk quality, productivity and behavior." *Annual Meeting – National Mastitis Council, Inc.* 15–24.

Leonardo, Bottaci, J. Drew Philip, E. Hartley John, B. Hadfield Matthew, and R. T. Monson John. 1997. "Artificial neural networks applied to outcome prediction for colorectal cancer patients in separate institutions." *The Lancet* 350: 469–472.

Lipton, Zachary Chase. 2016. "The Mythos of Model Interpretability." *CoRR*.

Lloyd, Blanchard, Zhao Bo, and Yinger John. 2005. *Do Credit Market Barriers Exist for Minority and Women Entrepreneurs?* Working Paper, Center for Policy Research.

Lyons, Samanthe M., Elaheh Alizadeh, Joshua Mannheimer, Katherine Schuamberg, Jordan Castle, Bryce Schroder, Philip Turk, Douglas Thamm, and Prasad Ashok. 2016. "Changes in cell shape are correlated with metastatic potential in murine and human osteosarcomas." *Biology Open* (The Company of Biologists Ltd) 5 (3): 289–299. https://bio.biologists.org/content/5/3/289.

Miller, Tim. 2017. "Explanation in Artificial Intelligence: Insights from the Social Sciences." *CoRR*.

Muhammad, Bilal Zafar, Isabel Valera, and Gomez Rodriguez, Gummadi, Krishna P. Manuel. 2015. "Fairness Constraints: Mechanisms for Fair Classification."

Narayanan, A., and V. Shmatikov. 2008. "Robust De-anonymization of Large Sparse Datasets." *2008 IEEE Symposium on Security and Privacy (sp 2008).* 111–125.

Nguyen, Barbara, Brady Werth, Nicholas Brewer, Jeanette G. Ward, R. Joseph Nold, and James M. Hann. 2017. "Comparisons of Medical Student Knowledge Regarding Life-Threatening CT Images Before and After Clinical Experience." *Kansas Journal of Medicine* (University of Kansas Medical Center) 10 (3): 1–12.

Oberman, Michelle. 2017. "The Sticky Standard of Care." *Hastings Center Report* 47 (5): 25–26.

OPEC. 2019. "OPEC Annual Statistical Bulletin 2019." OPEC.

Oxford Living Dictionaries. Accessed October 10, 2020. www.lexico.com/definition/artificial_intelligence.

Patatouka, E., and A. Fasianos. 2015. "Credit Discrimination in European Households – Evidence from survey data in Eurozone and the case of Greece." Germany.

Phong, Le Trieu, Yoshinori Aono, Takuya Hayashi, Lihua Wang, and Shiho Moriai. 2018. "Privacy-Preserving Deep Learning via Additively Homomorphic Encryption." *Trans. Info. For. Sec.* (IEEE Press) 13 (5): 1556–6013.

Ponikowski, Piotr, Stefan D. Anker, Khalid F. AlHabib, Martin R. Cowie, Thomas L. Force, Shengshou Hu, Tiny Jaarsma, et al. 2014. "Heart failure: preventing disease and death worldwide." *ESC Heart Failure* 4–25.

Rabin, Matthew. 2002. "Inference by Believers in the Law of Small Numbers." *The Quarterly Journal of Economics* (Oxford University Press) 117 (3): 775–816. www.jstor.org/stable/4132489.

Rafique, Muhammad, Raza Tunio, Imran Shah, Shah Abdul, Khairpur Mir's, and Sindh. 2017. "Factors Affecting to Employee's Performance. A Study of Islamic Banks." *International Journal of Academic Research in Accounting, Finance and Management Sciences*.

Raman, Rajiv, Padmaja Kumari Rani, Sudhir Reddi Rachepalle, Perumal Gnanamoorthy, Satagopan Uthra, Govindasamy Kumaramanickavel, and Tarun Sharma. 2009. "Prevalence of diabetic retinopathy in India: Sankara Nethralaya Diabetic Retinopathy Epidemiology and Molecular Genetics Study report 2." *Ophthalmology* 311–318.

Rusu, Gabriela, Silvia Avasilcai, and A.C. Huţu. 2016. "Organizational Context Factors Influencing Employee Performance Appraisal: A Research Framework." *Procedia – Social and Behavioral Sciences* 221: 57–65.

Saadi, Lahlou. 2018. "The Structure of Installations." In *Installation Theory: The Societal Construction and Regulation of Behaviour*, 93–174. Cambridge: Cambridge University Press.

Sackett, DL, WM Rosenberg, JA Gray, RB Haynes, and WS Richardson. 1996. "Evidence based medicine: what it is and what it isn't." *BMJ (Clinical research ed.)* 312 (7023): 71–72.

Salkowski, Lonie R., and Rosemary Russ. 2018. "Cognitive processing differences of experts and novices when correlating anatomy and cross-sectional imaging." *Journal of Medical Imaging* (Society of Photo-Optical Instrumentation Engineers) 5 (3).

Savage, Neil. 2016. "Privacy: The myth of anonymity." *Nature*. Accessed March 16, 2021. www.nature.com/articles/537S70a.

Scherer, Matthew U. 2015. "Regulating Artificial Intelligence Systems: Risks, Challenges, Competencies, and Strategies." *Harvard Journal of Law & Technology* 29 (2).

Shanteau, James, and Alleene Pingenot. 2009. "Subjective Expected Utility Theory." In *M. W. Kattan (Ed.), Encyclopedia of Medical Decision Making*. Sage Publications, Inc.

Shearer, Colin. 2000. "The CRISP-DM model: the new blueprint for data mining." *Journal of Data Warehousing* 5 (4): 13–22.

Shokri, Reza, and Vitaly Shmatikov. 2015. "Privacy-Preserving Deep Learning." *Proceedings of the 22nd ACM SIGSAC Conference on Computer and Communications Security*. Denver, Colorado, USA: ACM. 1310–1321.

Simon, Herbert A. 1957. "Models of Man: Social and Rational." *The Economic Journal* 279.

Simpkin, Arabella L., and Richard M. Schwartzstein. 2016. "Tolerating uncertainty – the next medical revolution?" *New England Journal of Medicine* (Massachusetts Medical Society) 375 (18).

Solomon, Y. Deku, Kara Alper, and Molyneux Philip. 2013. *Exclusion and Discrimination in the Market for Consumer Credit*. Working paper, Centre for Responsible Banking and Finance (RBF) of the University of St Andrews School of Management.

Stefan, M., F. Holzmeister, A. Müllauer, and M. Kirchler. 2018. "Ethnical discrimination in Europe: Field evidence from the finance industry." *PLoS ONE*.

Swire, Peter. 1997. "Markets, Self-Regulation, and Government Enforcement in the Protection of Personal Information, in Privacy and Self-Regulation in the Information Age by the U.S. Department of Commerce." SSRN.

Tang, Fengyi, Wei Wu, Jian Liu, Huimei Wang, and Ming Xian. 2019. "Privacy-Preserving Distributed Deep Learning via Homomorphic Re-Encryption." *Electronics*.

Thomas, N. Kipf, and Max Welling. 2016. "Semi-Supervised Classification with Graph Convolutional Networks." *CoRR*.

UK Department of Health and Social Care. 2021. "Vaccinations for coronavirus." *gov.uk*. February 19. Accessed March 12, 2021. www.gov.uk/government/news/new-campaign-to-support-vaccine-roll-out-backed-by-social-media-companies-and-british-institutions.

Vasileios, Iosifidis, and Ntoutsi Eirini. 2018. "Dealing with Bias via Data Augmentation in Supervised Learning." *Proceedings of the International Workshop on Bias in Information, Algorithms, and Systems (BIAS).* Sheffield, United Kingdom. 24–29.

Weiss, Sholom, Nitin Indurkhya, Tong Zhang, and Fred Damerau. 2004. *Text Mining: Predictive Methods for Analyzing Unstructured Information.* Springer Verlag.

WhatsApp LLC. 2021. *Answering your questions about WhatsApp's Privacy Policy.* 02. Accessed March 06, 2021. https://faq.whatsapp.com/general/security-and-privacy/answering-your-questions-about-whatsapps-privacy-policy.

Whittingham, Robert B. 2008. *Preventing corporate accidents: an ethical approach.* Amsterdam; Boston, Mass.: Butterworth-Heinemann.

Wilkinson, C. P., Frederick L. 3rd Ferris, Ronald E. Klein, Paul P. Lee, Carl David Agardh, Matthew Davis, Diana Dills, Anselm Kampik, R. Pararajasegaram, and Juan T. Verdaguer. 2003. "Proposed international clinical diabetic retinopathy and diabetic macular edema disease severity scales." *Ophthalmology* 1677–1682.

William, C. Shiel Jr. 2021. "standard_of_care/definition.htm." *www.medicinenet.com.* March 28. Accessed March 28, 2021. www.medicinenet.com/standard_of_care/definition.htm.

World Health Organization. 2021. *Health Workforce.* Accessed March 28, 2021. www.who.int/data/gho/data/themes/topics/health-workforce.

I

Index

A

Adversarial machine learning, 60

AI and regulations
  customers, regulatory arrangement, 92–94
  data protection laws, 84
  data related revenue
    data, 94, 95
    data sharing, 99, 100
    data tax, 95–97
    mechanisms, 101
  DS, 86–89
  ethical behavior and compliance, 85
  GDPR, 84
  reasons, 84
  regulators, 84
  third party regulatory arrangement, 90, 91

AI-assisted services, 125

AI-based CDSS system, 113

AI performance metrics
  ML algorithms, 46
  privacy, 46
  supervised algorithm, 46
  supervised problems, 46
    AUC, 49–51
    choosing right, 51, 52
    classification accuracy, 47
    confusion matrix, 48, 49
    credit card fraud, 47
    DS team, 47
    features, 54
    F-measure, 52, 53

    fraudulent/non-fraudulent, 47
    logarithmic loss, 48
    regression problems, 53, 54
  system performance, 46
  unsupervised algorithm, 46
  unsupervised problems, 46
    applications, 56
    association analysis, 55, 56
    clustering, 54, 55
    data visualization, 55
    dimension reduction, 55
    rules, 56

Amazon, 4

Application programming interface (API), 61

Area under curve (AUC), 47, 49–51

Artificial intelligence (AI), 123
  accuracy, 45
  agents/systems, 4
  challenges, 6, 7
  data sets, 1
  definitions, 2–4
  development/deployment, 28
  DS practitioners, 45
  elements, 14
  internet, 1
  mainstream public perception, 27
  organizations, 4, 5
  sensitive information, 14
  social norms, 3
  sustainable
    DS, 7
    governance, 9

© Ghislain Landry Tsafack Chetsa 2021
G. L. Tsafack Chetsa, *Towards Sustainable Artificial Intelligence*,
https://doi.org/10.1007/978-1-4842-7214-5

Printed in the United States
by Baker & Taylor Publisher Services